MATHS PLUS

2

MENTALS AND HOMEWORK BOOK

AUSTRALIAN CURRICULUM

Harry O'Brien
Greg Purcell

OXFORD
UNIVERSITY PRESS

Contents

Contents

Australian Curriculum Linking Chart

Units	1	2	3	4	5
Number and Algebra					
Recognise, represent and order numbers to at least 1000 using physical and virtual materials, numerals and number lines (AC9M2N01)	■				
Partition, rearrange, regroup and rename two- and three-digit numbers using standard and non-standard groupings; recognise the role of a zero digit in place value notation (AC9M2N02)	■			■	
Recognise and describe one-half as one of 2 equal parts of a whole and connect halves, quarters and eighths through repeated halving (AC9M2N03)			■		■
Add and subtract one- and two-digit numbers, representing problems using number sentences, and solve using part-part-whole reasoning and a variety of calculation strategies (AC9M2N04)	■	■	■	■	
Multiply and divide by one-digit numbers using repeated addition, equal grouping, arrays, and partitioning to support a variety of calculation strategies (AC9M2N05)					■
Use mathematical modelling to solve practical problems involving additive and multiplicative situations, including money transactions; represent situations and choose calculation strategies; interpret and communicate solutions in terms of the situation (AC9M2N06)	■	■	■	■	■
Recognise, describe and create additive patterns that increase or decrease by a constant amount, using numbers, shapes and objects, and identify missing elements in the pattern (AC9M2A01)					■
Recall and demonstrate proficiency with addition facts to 20; extend and apply facts to develop related subtraction facts (AC9M2A02)		■	■		
Recall and demonstrate proficiency with multiplication facts for twos; extend and apply facts to develop the related division facts using doubling and halving (AC9M2A03)			■		■
Measurement and Space					
Measure and compare objects based on length, capacity and mass using appropriate uniform informal units and smaller units for accuracy when necessary (AC9M2M01)	■				■
Identify common uses and represent halves, quarters and eighths in relation to shapes, objects and events (AC9M2M02)			■		■
Identify the date and determine the number of days between events using calendars (AC9M2M03)					
Recognise and read the time represented on an analog clock to the hour, half-hour and quarter-hour (AC9M2M04)		■		■	
Identify, describe and demonstrate quarter, half, three-quarter and full measures of turn in everyday situations (AC9M2M05)			■		
Recognise, compare and classify shapes, referencing the number of sides and using spatial terms such as "opposite", "parallel", "curved" and "straight" (AC9M2SP01)	■			■	
Locate positions in two-dimensional representations of a familiar space; move positions by following directions and pathways (AC9M2SP02)		■			
Statistics and Probability					
Acquire data for categorical variables through surveys, observation, experiment and using digital tools; sort data into relevant categories and display data using lists and tables (AC9M2ST01)					■
Create different graphical representations of data using software where appropriate; compare the different representations, identify and describe common and distinctive features in response to questions (AC9M2ST02)					■

6	7	8	9	10	11	12	13	14	15	16	17	18	19	20	21	22	23	24	25	26	27	28	29	30	31	32

Number and Algebra

SET 1 Basic

1 5 + 1 = ☐

2 4 + 1 = ☐

3 3 + 1 = ☐

4 4 − 1 = ☐

5 6 + 1 = ☐

6 8 + 1 = ☐

7 2 + 1 = ☐

8 1 + 1 = ☐

9 7 − 1 = ☐

10 If Tim has 3 cards and Mark has 4, how many do they have altogether? ☐

SET 2 Numbers to 100

Complete the number expanders.

1 24 = ☐ tens ☐ ones

2 32 = ☐ tens ☐ ones

3 56 = ☐ tens ☐ ones

4 12 = ☐ tens ☐ ones

5 28 = ☐ tens ☐ ones

6 99 = ☐ tens ☐ ones

Calculate the answers.

7 ☐ = 6 tens 2 ones

8 ☐ = 7 tens 5 ones

Space Rectangles and squares

1 Put a cross on all rectangles.

2 Put a dot on all squares.

3 Put a tick on all triangles.

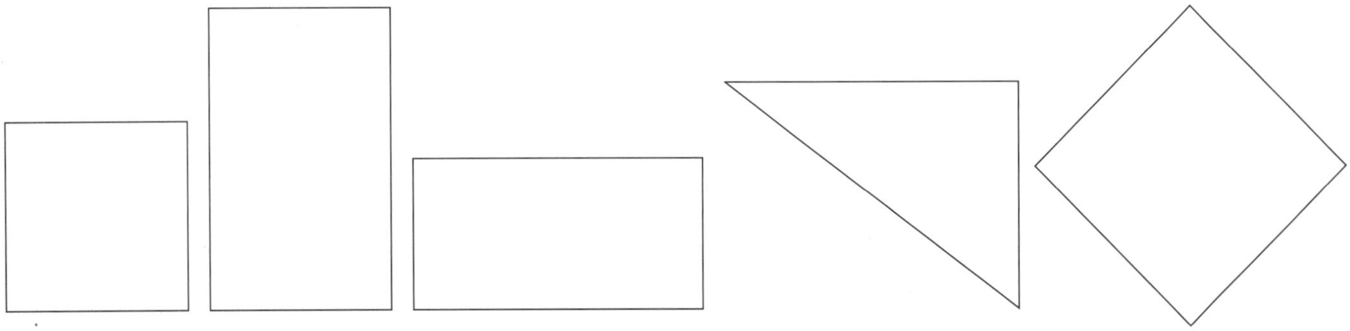

Number and Algebra

SET 3 Addition facts to 20

1 Colour in the ten frames to complete the additions. Complete the answers.

a

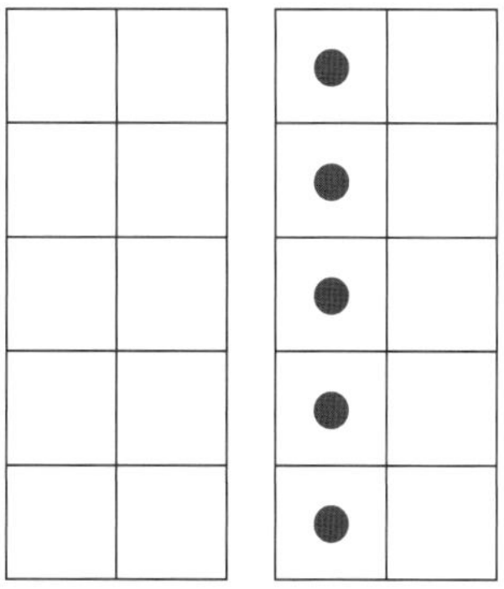

☐ + 5 = 10

b

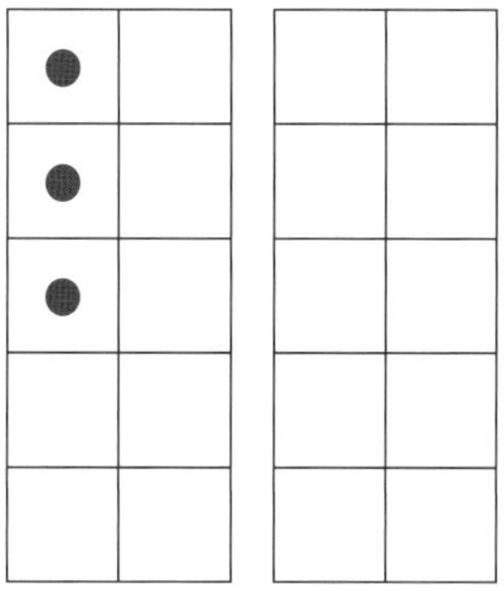

3 + ☐ = 6

Complete the Base 10 addition grid.

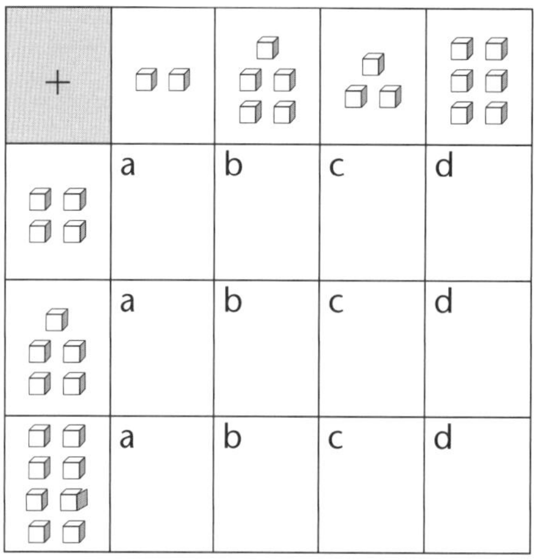

SET 4 Extension

1 25 = ☐ tens + ☐ ones

2 Which is smaller: 9 + 3 or 4 + 6?

3 How many sides does a square have?

4 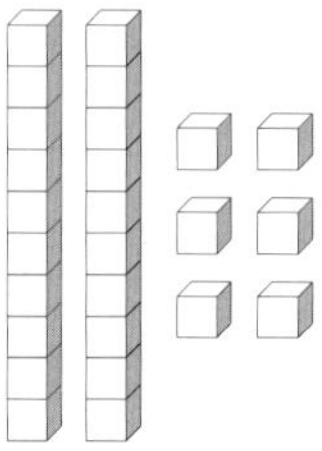 and 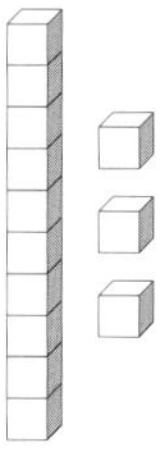 makes ☐

5 What is the sum of 9 and 7?

6 8 tens + 2 ones = ☐

7 Add $6 to $4.

8 Which number is bigger: 10 + 3 or 9 + 9?

9 Shade two numbers in each circle that total 12.

a

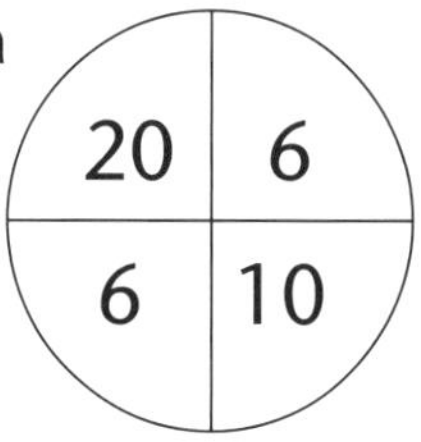

b

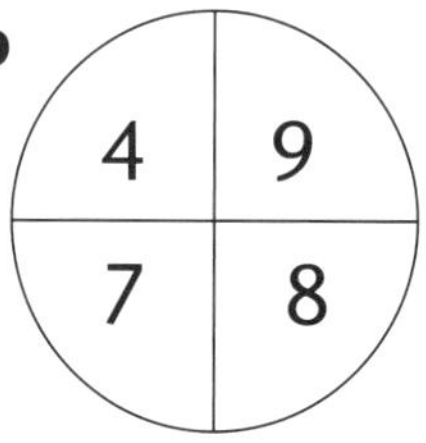

Measurement Length

1 Colour the two pencils that are the same length yellow.

2 Colour the pencil that is the shortest red.

3 Colour the pencil that is the longest blue.

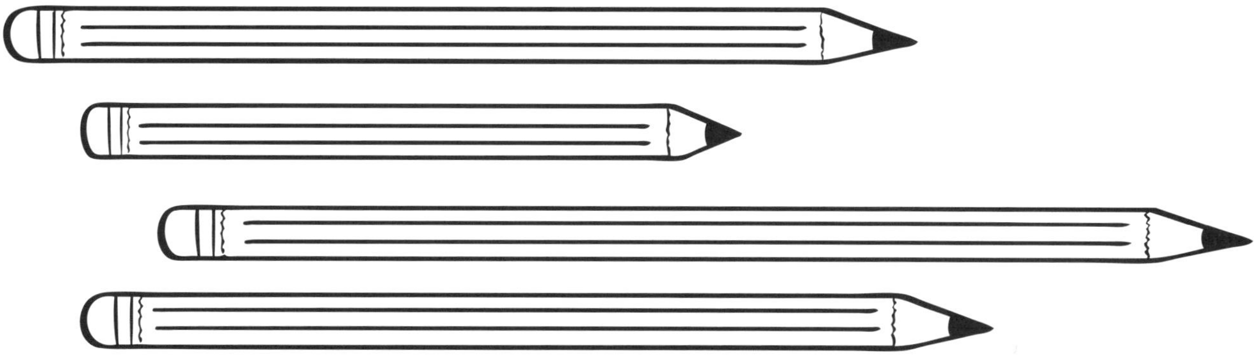

Number and Algebra

SET 1 Basic

		+ 1
1	2	
2	4	
3	6	
4	7	
5	5	
6	8	

7 8 + 0 = ☐

8 6 − 1 = ☐

9 4 − 1 = ☐

10

Eve baked 6 cakes but ate one. How many are left?

☐

SET 2 Subtraction

Write true or false to describe each number sentence. Use the stars to help you count.

1 9 − 4 = 8 False

2 10 − 3 = 6 ☐

3 15 − 6 = 9 ☐

4 12 − 6 = 6 ☐

5 7 − 4 = 3 ☐

6 8 − 2 = 4 ☐

7 19 − 1 = 17 ☐

8 10 − 8 = 2 ☐

Space Describing position

Use the words to help you answer the questions.

piggy bank, clock, male doll, aeroplane, glove, tin soldier, car

1 Which item is above the clock? ______________

2 Which item is below the teddy? ______________

3 Which item is beside the piggy bank? ______________

4 Which item is in the bottom left? ______________

5 Which item is in the top right? ______________

Number and Algebra

SET 3 Count on/addition

Use the strategy outlined or use any other strategy to answer the questions.

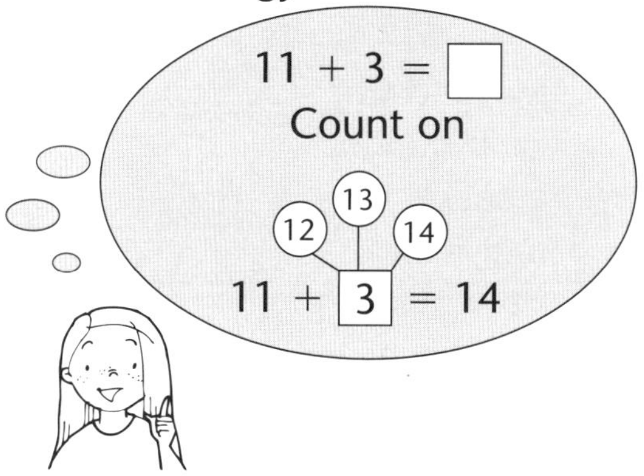

1. 6 + 8 = ☐
2. 10 + 6 = ☐
3. 11 + 8 = ☐
4. 12 + 10 = ☐
5. 7 + 3 = ☐
6. 6 + 6 = ☐
7. 10 + 8 = ☐
8. 15 + 7 = ☐
9. 4 + 9 = ☐

SET 4 Extension

1. 8 + 8 = ☐
2. Which number is smallest: 267, 672 or 126?
3. 8 − 6 = ☐
4. How many 10c coins make 50c?
5. Write the number thirty-six. ☐
6. 56 = ☐ tens ☐ ones
7. What is the difference between $17 and $7?
8. What number is between 19 and 21?
9. Complete: 1st, 2nd, 3rd, ____, ____
10. Write three number sentences that have an answer of 5.

a ☐ + ☐

b ☐ + ☐ = 5

c ☐ + ☐

Measurement Time

What time is it?

1

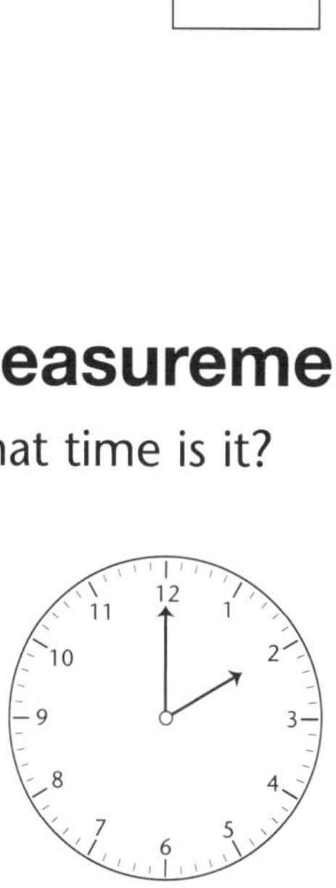

☐ o'clock

2

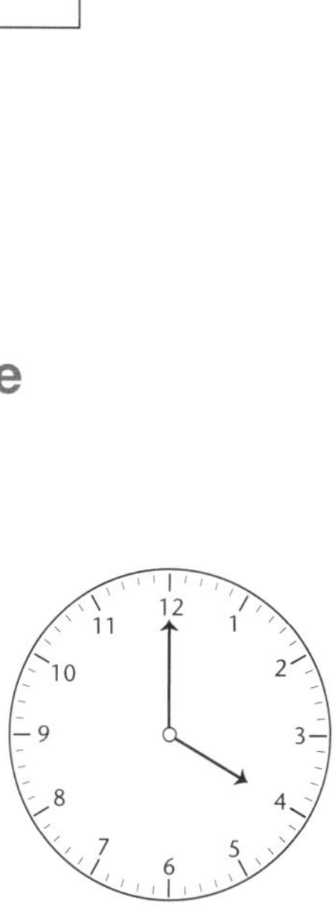

☐ o'clock

3

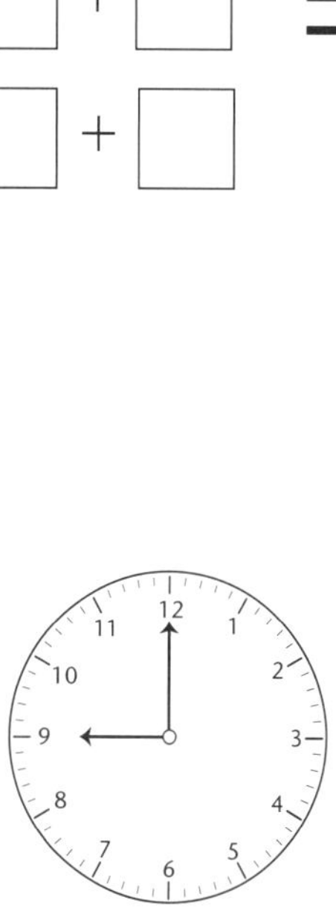

☐ o'clock

4

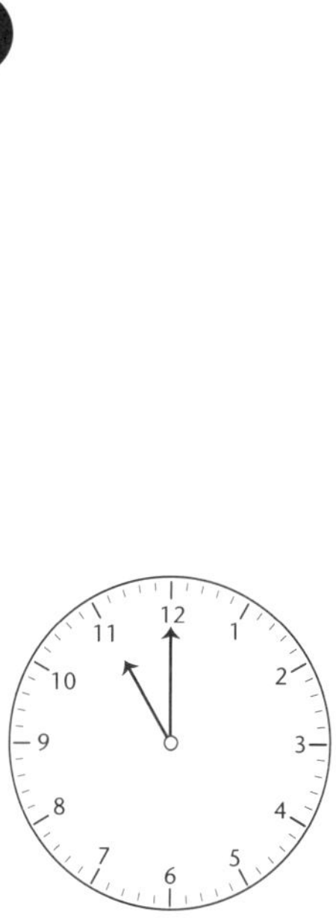

☐ o'clock

Number and Algebra

SET 1 Basic

1 2 + 4 = ☐

2 3 + 2 = ☐

3 3 − 2 = ☐

4 4 + 3 = ☐

5 5 − 1 = ☐

6 4 − 2 = ☐

7 2 + 2 = ☐

8 6 − 2 = ☐

9 7 + 1 = ☐

10 Joel and Chris each bought a packet of four lollies. How many do they have between them? ☐

SET 2 Addition to 20

Finish the number sentences.

1 7 + 2 = ☐

2 6 + 3 = ☐

3 8 + 2 = ☐

4 5 + 4 = ☐

5 9 + 1 = ☐

6 7 + 6 = ☐

7 3 + 8 = ☐

8 9 + 3 = ☐

9 12 + 2 = ☐

10 13 + 4 = ☐

11 11 + 5 = ☐

Number and Algebra Doubling and halving

In the second box, draw double the number of items you can count in the first box.

First box	Second box
☐ ☐ ☐	

double 3 is ☐

half of 6 is ☐

Number and Algebra

SET 3 Halves

Shade one half of each shape or group.

1 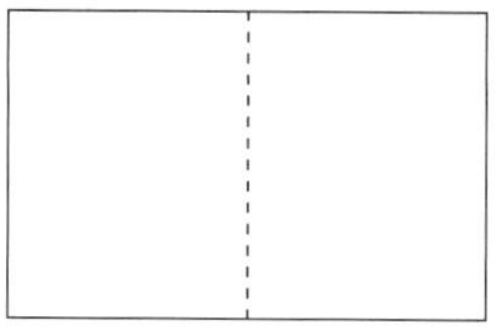

2 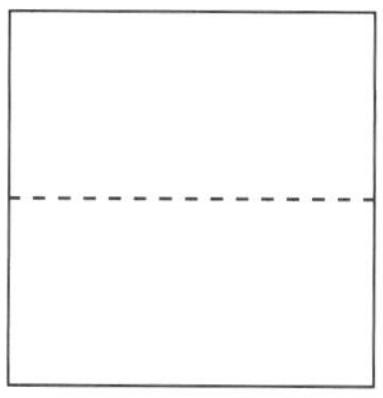

3 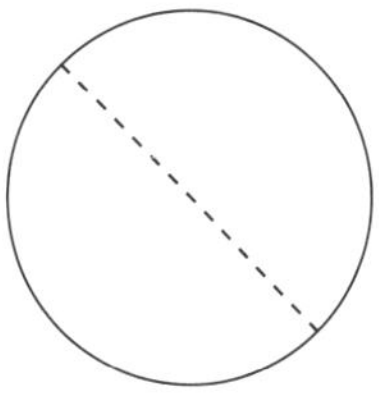

4 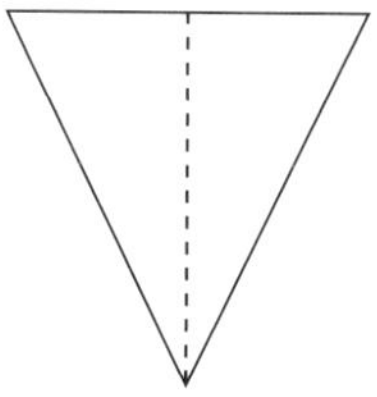

5 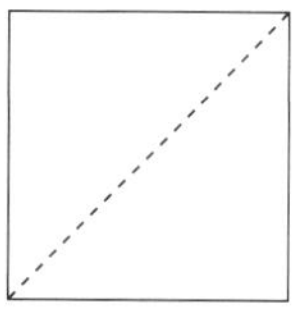

6 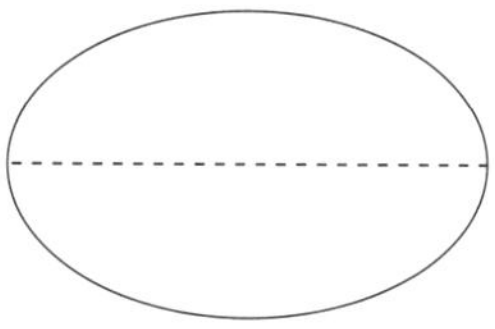

7 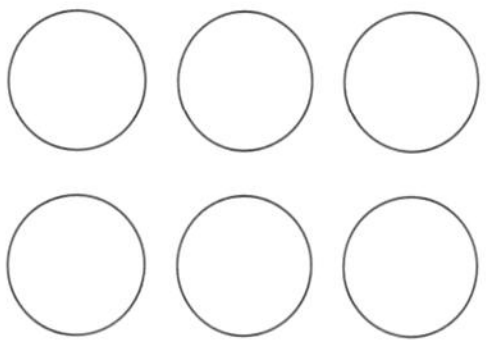

8 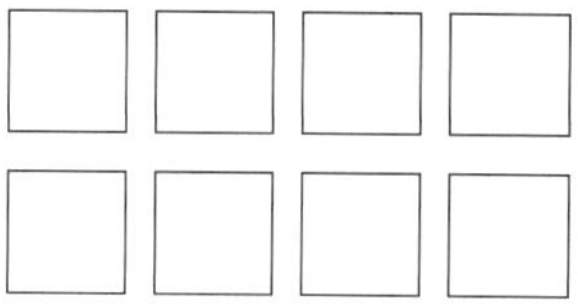

SET 4 Extension

1 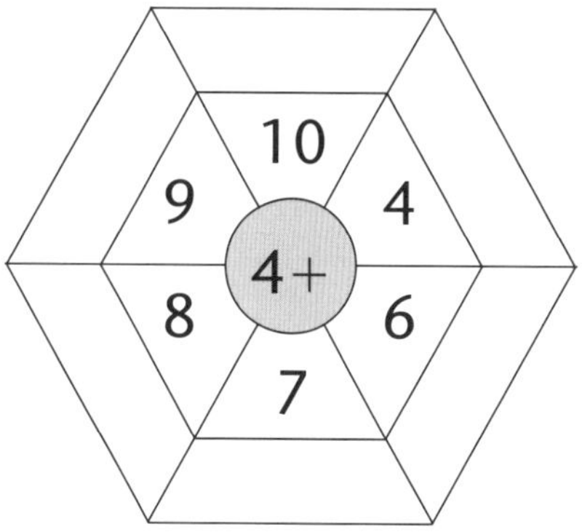

2 Which season comes after autumn?

3 13 + 4 = ☐

4 Subtract 5 from 20.

5 7 + 10 = 17, 17 − 7 = ☐

6 8 × 0

7 What number is between 30 and 32?

8 18 = ☐ tens ☐ ones

9 By how much is 10 less than 17?

10 Circle the smallest number: 215, 512, 125

11 What month is between March and May?

Measurement Clockwise and anticlockwise turns

Turn each letter clockwise for a quarter turn.

1 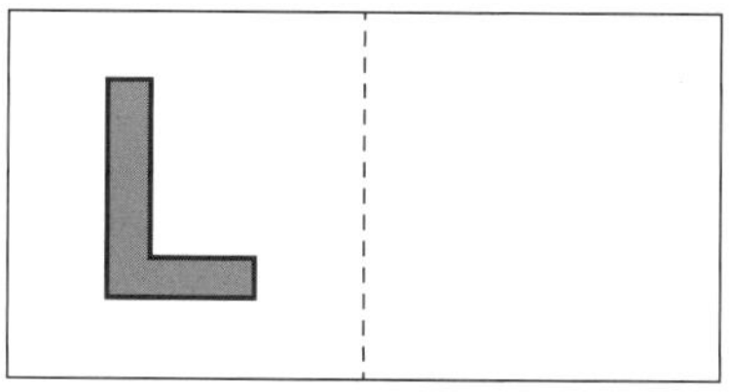

2 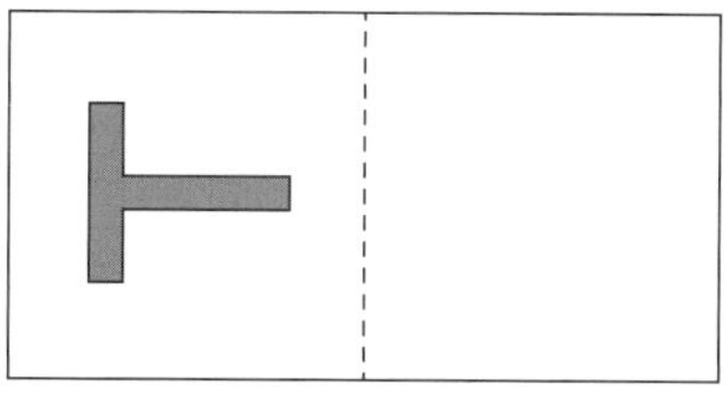

3 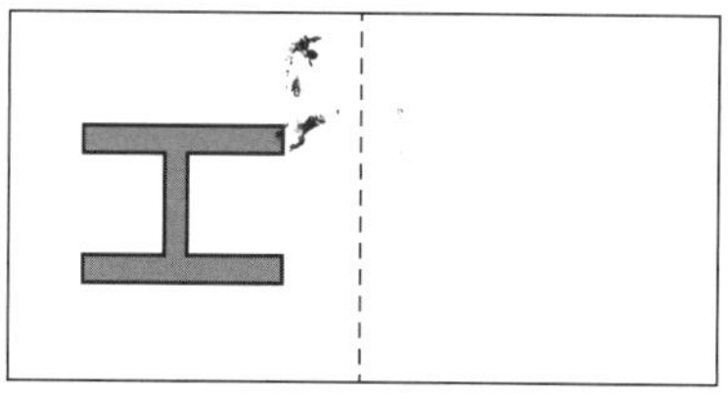

Turn these triangles anticlockwise for a quarter turn.

4 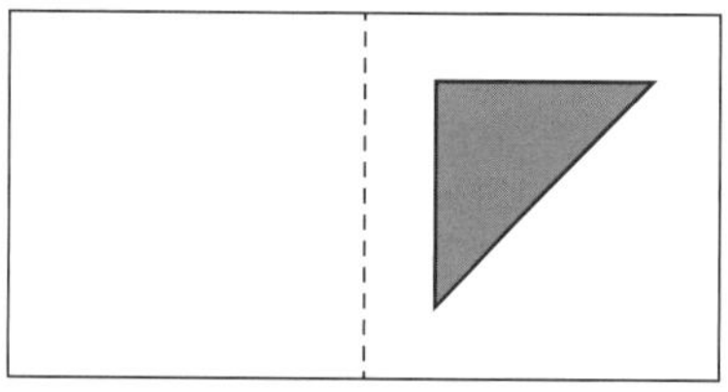

5 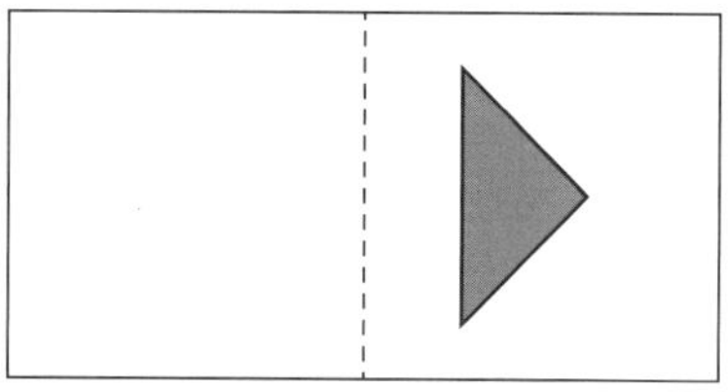

6 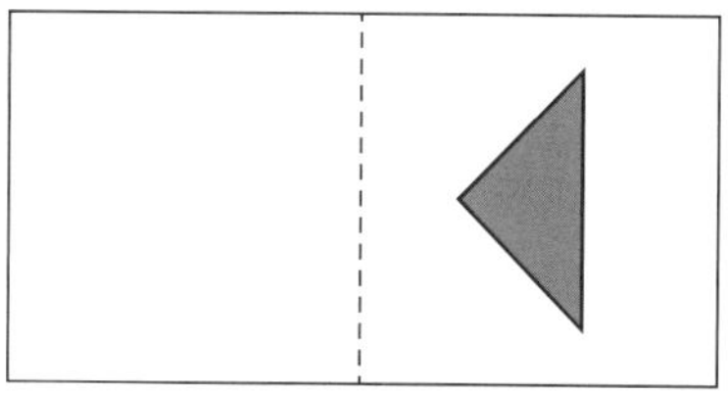

Number and Algebra

SET 1 Basic

1 5 + 1 = ☐

2 4 + 4 = ☐

3 5 − 2 = ☐

4 5 + 3 = ☐

5 8 − 3 = ☐

6 6 + 2 = ☐

7 4 − 2 = ☐

8 5 − 4 = ☐

9 7 − 2 = ☐

10

What is the number 3 doubled?

☐

SET 2 Subtraction facts

Find the missing numbers in these subtractions.

1 10 − ☐ = 5

2 4 − ☐ = 2

3 6 − ☐ = 3

4 ☐ − 10 = 15

5 ☐ − 2 = 10

6 ☐ − 3 = 7

7 10 − ☐ = 3

8 12 − ☐ = 10

9 8 − ☐ = 4

Space Recognising 3D objects

Join the dots to form the objects, then name them.

1

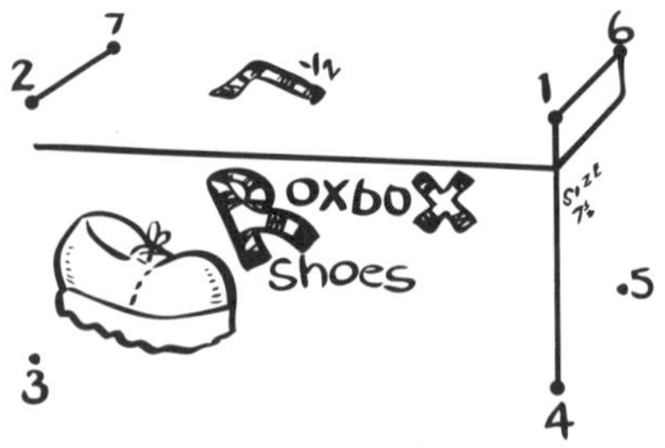

☐

2

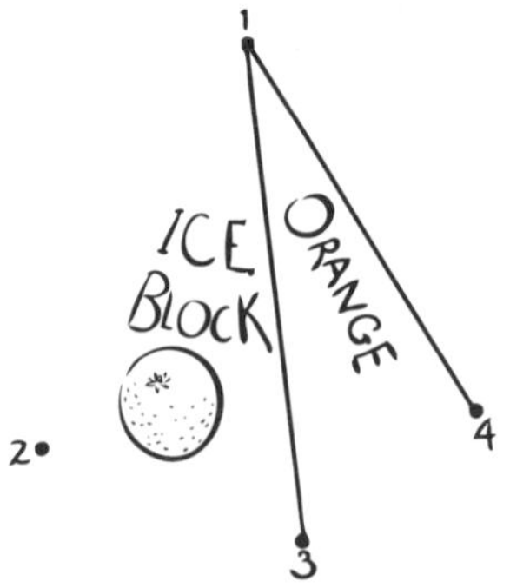

☐

3

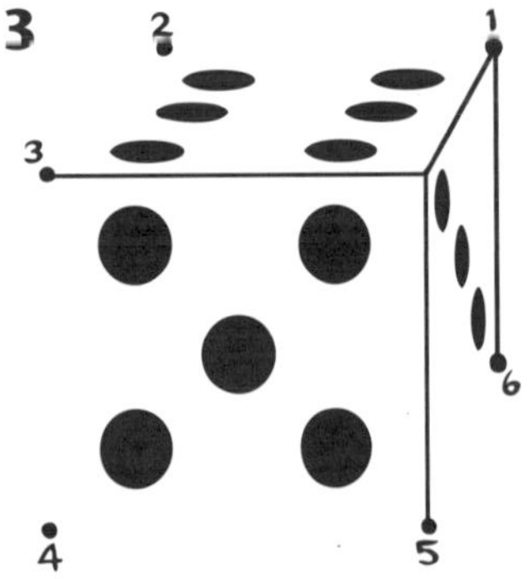

☐

Number and Algebra

SET 3 Place value and counting over 100

1

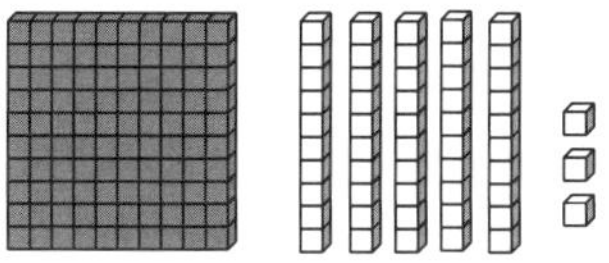

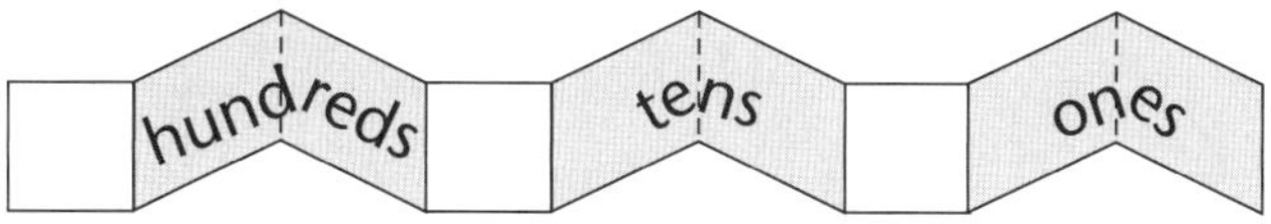

2

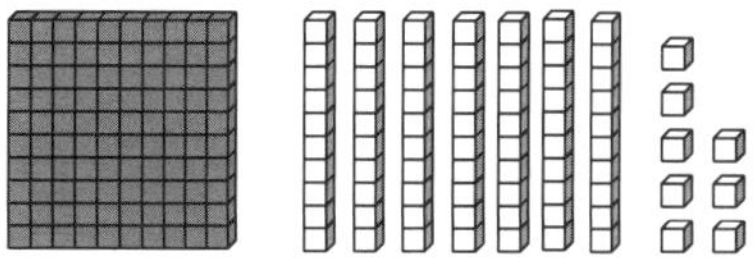

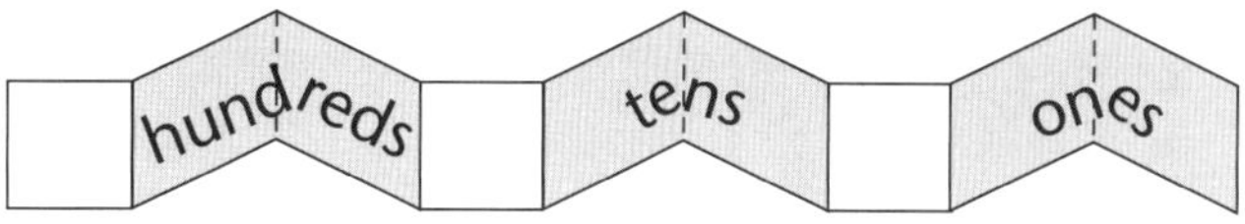

SET 4 Extension

1 Write the word for 7.

2 How many 10c coins make $1?

3 How many eggs in one dozen?

4 Is 72 odd or even?

5 Which is longer: 1 m or 100 cm?

6 How many days in 2 weeks?

7 A dozen – 5

8 Half of 16

9 6 × 5

10 What month comes before November?

11 How much money is here?

Measurement Time

Draw the times on the clock faces.

1

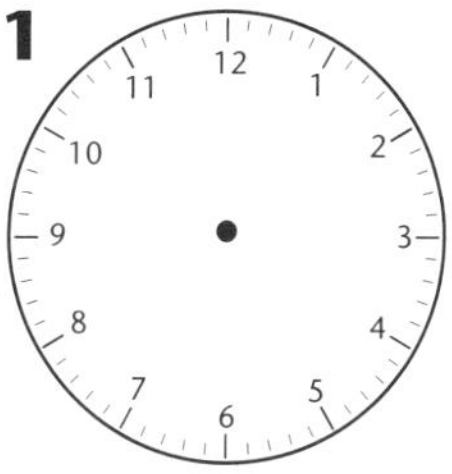

5 o'clock

2

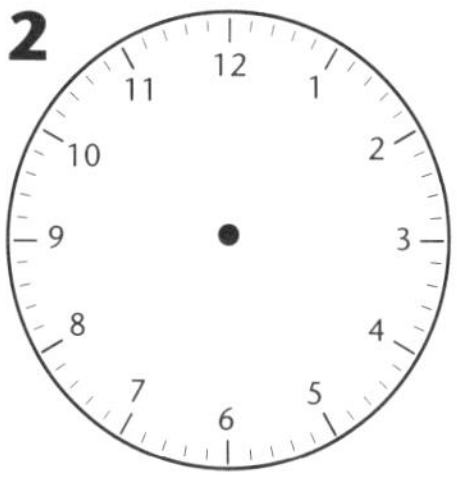

Half past 3

3

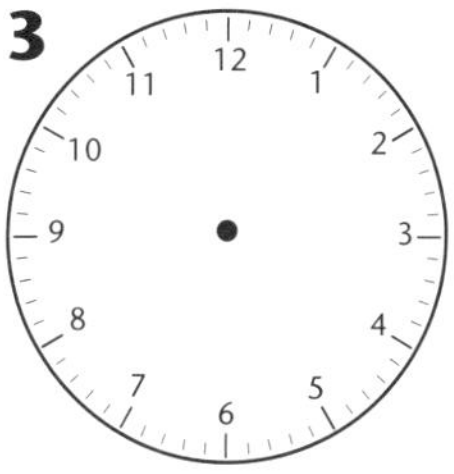

Half past 9

4

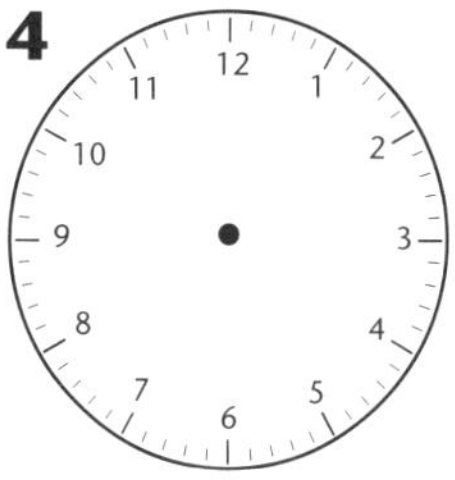

Half past 7

5

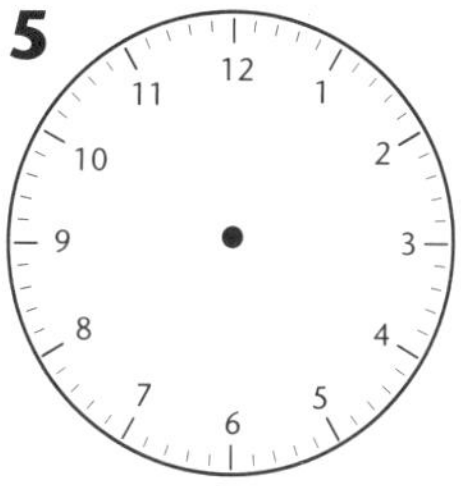

Half past 10

Number and Algebra

SET 1 Basic

		−1
1	2	
2	3	
3	4	
4	5	
5	6	
6	7	

7 [4] + [1] = [ ]

8 [8] − [1] = [ ]

9 [6] + [1] = [ ]

10 What number comes after six? [ ]

SET 2 Counting by 2s, 4s and 5s

Complete the number patterns by counting in 2s, 4s or 5s.

1 [2] [4] [6] [ ] [ ]

2 [5] [10] [15] [ ] [ ]

3 [14] [18] [22] [ ] [ ]

4 [4] [8] [12] [ ] [ ]

5 [20] [25] [30] [ ] [ ]

6 [22] [26] [30] [ ] [ ]

7 [70] [65] [60] [ ] [ ]

8 [60] [58] [56] [ ] [ ]

Measurement Length

1 How many shoes will fit along the table?

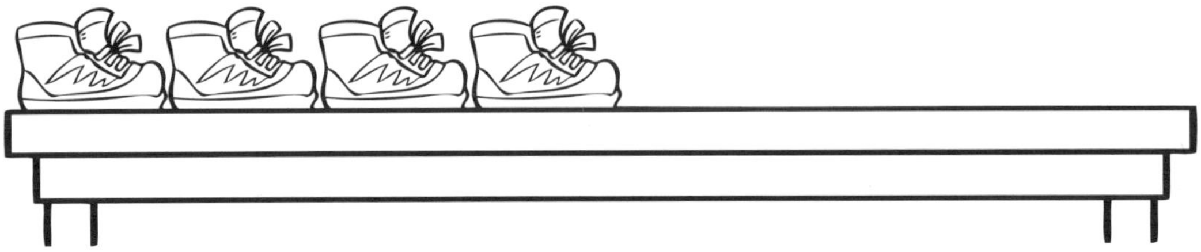

2 How many shoe boxes will fit along the table?

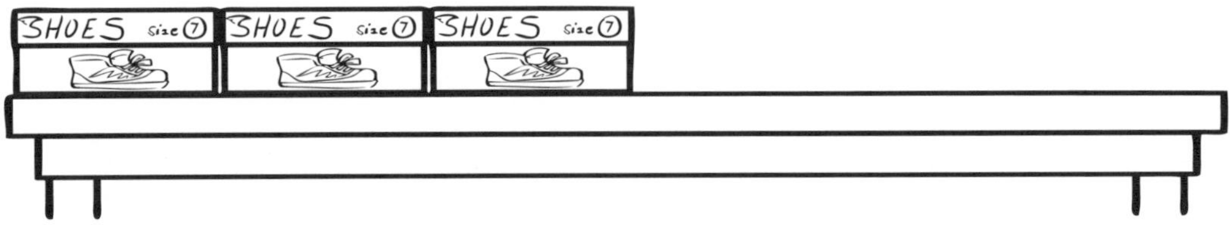

Number and Algebra

SET 3 One quarter

1 Shade $\frac{1}{4}$ of each shape or group.

a
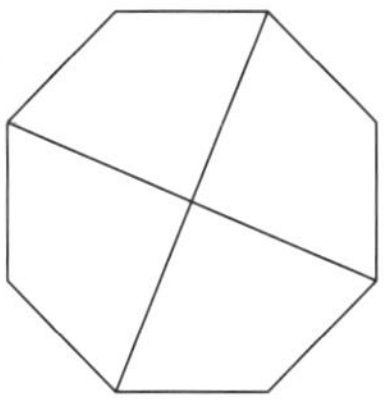

b
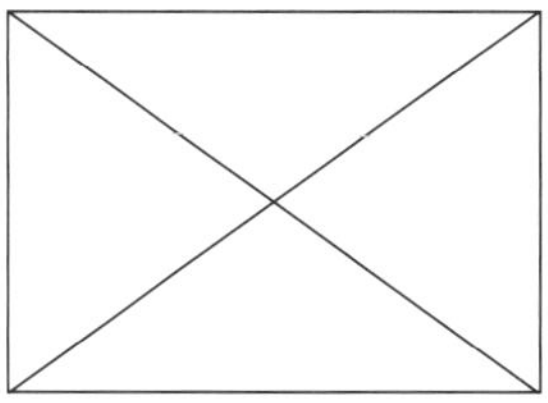

c
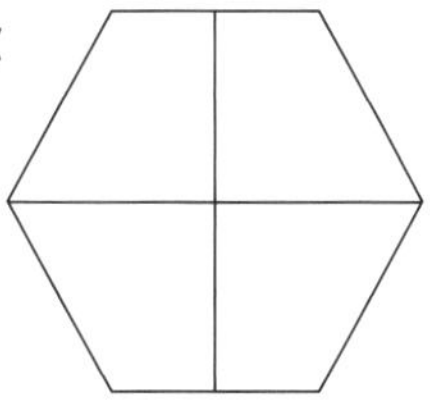

d
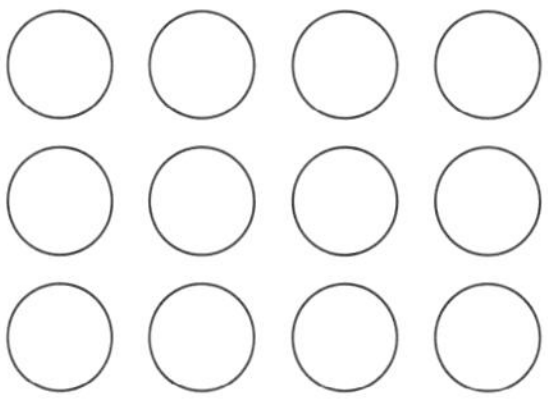

2 Circle $\frac{1}{4}$ of each group.

a

b

c
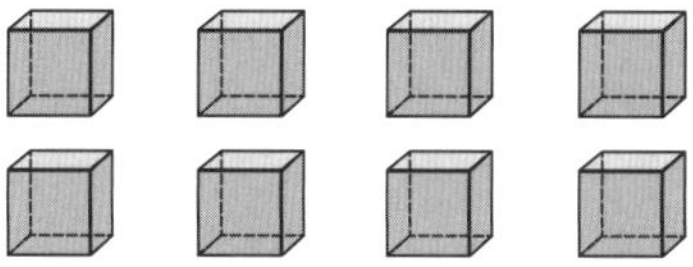

SET 4 Extension

1 What number is three more than 28?

2 10 + _____ = 20

3
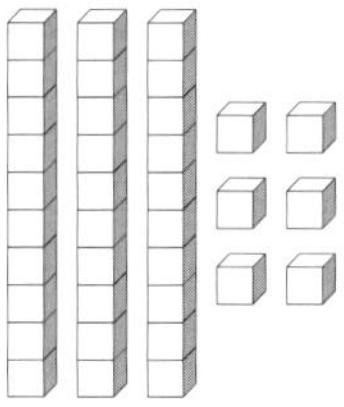
= ☐ tens ☐ ones

4 What month comes after November?

5 7 + 6 =

6 Jase had 13 books but dropped 3. How many does he still have?

7 10 + 11 =

8 Draw a rectangle.

9 8 − ☐ = 6

10 356 = 3 hundreds ☐ tens ☐ ones

11 Colour in the 2D square.

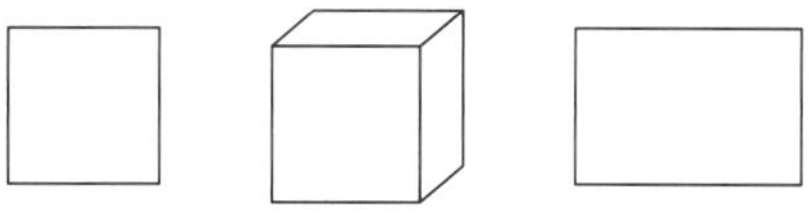

12 How many months in a year?

Statistics Tally marks

1 Make a tally of each pet. (𝍸)

Cat		**Bird**	
Dog		**Lizard**	
Fish			

2 Colour the blocks to make a column graph of your tally.

Favourite Pets

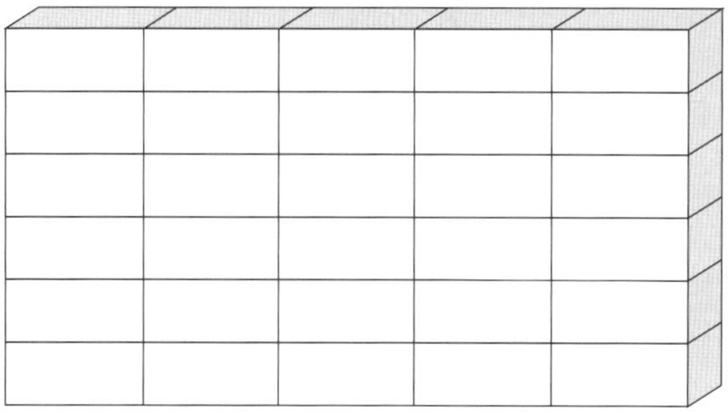

Number and Algebra

SET 1 Basic

1 5 + 5 = ☐

2 6 + 3 = ☐

3 8 + 1 = ☐

4 7 − 2 = ☐

5 10 + 2 = ☐

6 8 − 3 = ☐

7 6 + 1 = ☐

8 7 − 2 = ☐

9 5 − 3 = ☐

10

SET 2 Repeated addition

Use repeated addition to count the groups of 10 or 5.

1

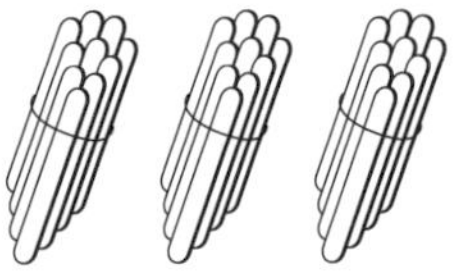

10 + ☐ + ☐ = ☐

2

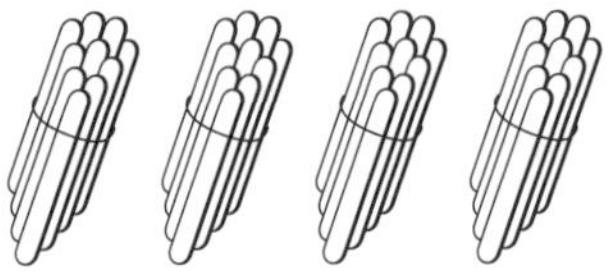

☐ + ☐ + ☐ + ☐ = ☐

3

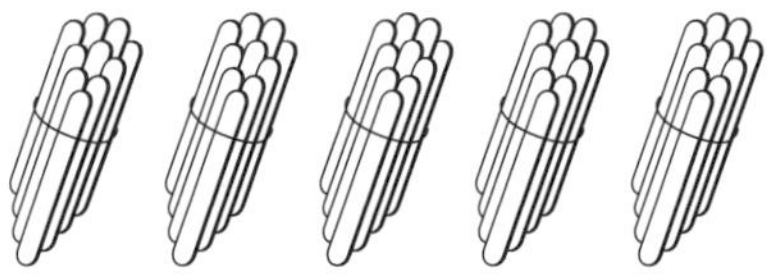

☐ + ☐ + ☐ + ☐ + ☐ = ☐

4

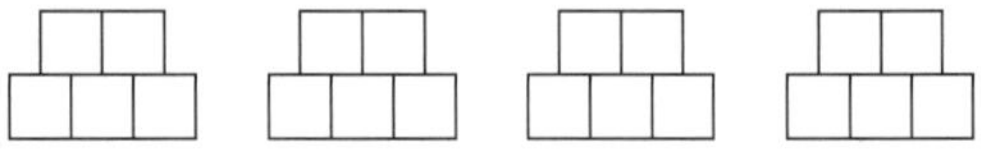

☐ + ☐ + ☐ + ☐ = ☐

Space Drawing 2D shapes

1 Construct a square that touches 4 dots.

2 Construct a rectangle that touches 8 dots.

3 Construct a rectangle that touches 10 dots.

4 Construct a square that touches 12 dots.

Number and Algebra

SET 3 Subtraction from addition

7 + 6 = 13
13 − 6 = 7

Complete each set of facts. The first two are done for you.

1. [8] + [4] = [12]
2. [12] − [4] = [8]
3. [14] + [ ] = [17]
4. [17] − [ ] = [14]
5. [10] + [ ] = [16]
6. [16] − [ ] = [10]
7. [6] + [ ] = [11]
8. [11] − [ ] = [6]
9. [2] + [ ] = [12]
10. [12] − [ ] = [2]

SET 4 Extension

1. Write a number sentence for the number 16.
   [ ] + [ ] = [16]
2. 20, 22, 24, ____, ____
3. Nine take away 3
4. What season comes after summer?
5. [3 + 7 = 10] ▲ [ ] + [ ] = 10
6. 13 − 4
7. 6 + [ ] = 9
8. 20c − 5c =
9. Which is heavier:
   1 kg of feathers or 1 kg of stone?
10. 20c − 15c =
11. [ ] − 2 = 18
12. What date is Anzac Day?
13. How many sides on one dice?

Measurement Capacity

Class T2 measured the capacity of 5 containers using a small plastic cup. They tallied their information.

1. Colour red the vessel that holds the most.
2. Colour blue the vessel that holds the least.

Juice	𝍸 II
Bottle	IIII
Jug	𝍸 𝍸 III
Mug	II
Vase	𝍸 I

UNIT 7

Number and Algebra

SET 1 Basic

		+2
1	2	
2	3	
3	4	
4	5	
5	6	
6	7	

7 [7] − [2] = [ ]

8 [8] + [2] = [ ]

9 [10] − [2] = [ ]

10 Jen bought 6 dresses but gave her friend Kelly 2. How many does Jen have left? [ ]

SET 2 Counting back

Count back to find the answer.

1 [14] − [ ] = [10]

2 [42] − [ ] = [38]

3 [17] − [ ] = [11]

4 [12] − [ ] = [8]

5 [21] − [ ] = [19]

6 [50] − [ ] = [45]

7 [35] − [ ] = [30]

8 [28] − [ ] = [22]

9 [46] − [ ] = [42]

Space Position

Pretend you are looking at the house, and the objects below.

1 Draw a tree to the **right** of the picture.

2 Draw a car to the **left** of the picture.

3 Draw a bird **on top of** the house.

4 Put some flowers **in front of** the house.

5 Draw an electricity pole **behind** the house. (You won't see the bottom of the pole.)

Number and Algebra

SET 3 Dividing/equal groups

Use the groups of popsticks to answer the number sentences.

1

9 sticks make ☐ groups of 3

2

12 sticks make ☐ groups of 4

3

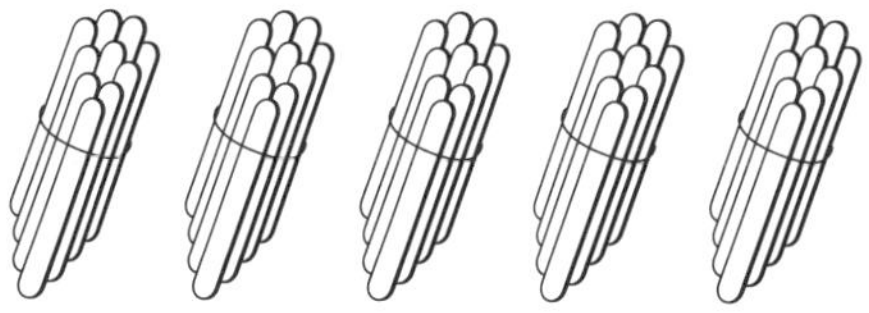

50 sticks make ☐ groups of 10

4

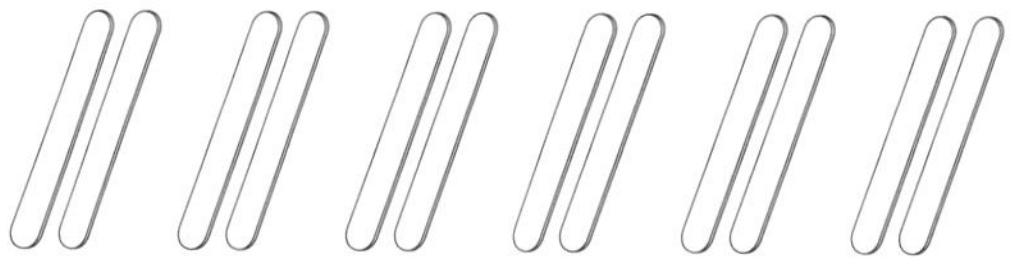

12 sticks make ☐ groups of 2

SET 4 Extension

1 Put these numbers in order from smallest to largest: 13, 10, 23, 99

2 $20 - \square = 6$

3 Half of 12

4 $7 + 7$

5 Add $12 to $8.

6 Is 32 an even number?

7 What is $1 + $0.65?

8 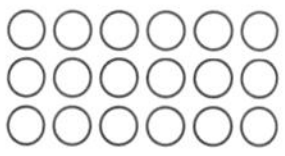 $18 \div 3 = \square$

9 What is 6 more than 14?

10 2×4

11 Write five number sentences that give the answer 20.

Measurement Mass/informal units

1 What could you use to balance this orange on the equal arm balance? ______

2 Describe how you would measure the mass of the orange. ______

______

______

______

______

______

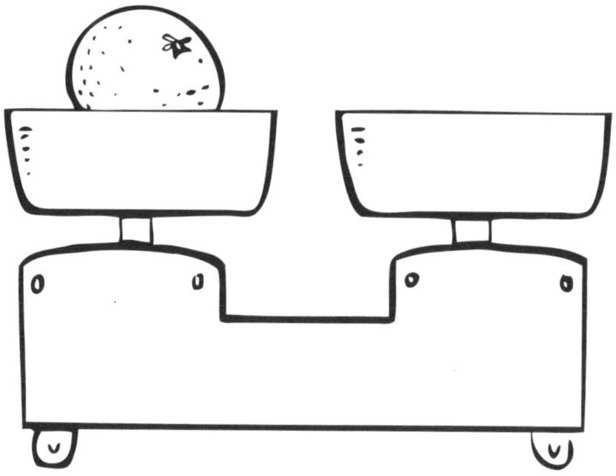

UNIT 8

Number and Algebra

SET 1 Basic

1 6 − 1 = ☐

2 2 + 2 = ☐

3 4 − 3 = ☐

4 10 − 6 = ☐

5 4 + 4 = ☐

6 3 + 2 = ☐

7 5 + 1 = ☐

8 7 + 1 = ☐

9 6 + 2 = ☐

10

One paddock had 3 sheep and another had 2. How many sheep are there altogether?

☐

SET 2 Multiplication/groups of

1

4 groups of 3 = ☐

2

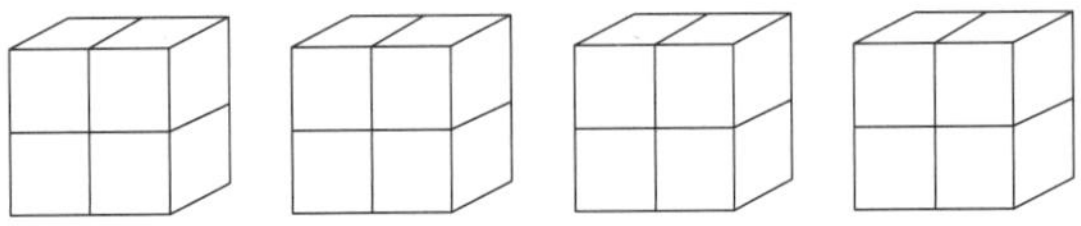

4 groups of 4 = ☐

3

5 groups of 2 = ☐

4

3 groups of 5 = ☐

Statistics Recording data

Use tally marks to record the number of marbles in the bag.

Marble	Tally
1	
2	
3	
4	
5	

Number and Algebra

SET 3 Subtraction from addition

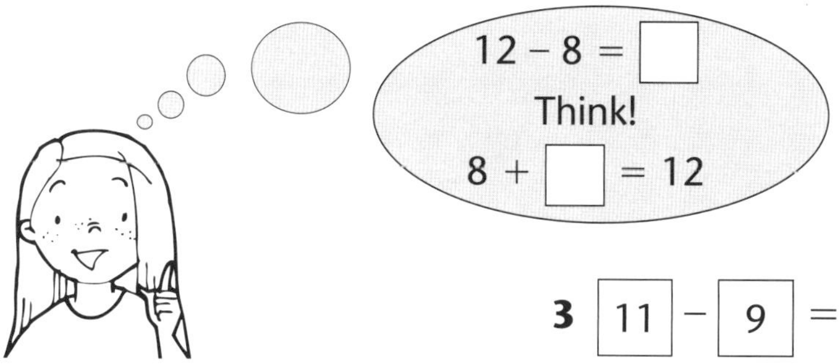

Use addition to solve the subtraction.

1. 9 – 7 = ☐
2. 12 – 9 = ☐
3. 11 – 9 = ☐
4. 12 – 8 = ☐
5. 13 – 10 = ☐
6. 14 – 8 = ☐
7. 15 – 11 = ☐
8. 16 – 12 = ☐
9. 17 – 14 = ☐
10. 18 – 15 = ☐

SET 4 Extension

1. Write the words for 310.
2. 6 + 2 + 1 = ☐
3. ☐ + 3 + 1 = 8
4. 15 – 0 =
5. How many wheels are on 2 cars?
6. Colour in $\frac{1}{4}$ of each shape.

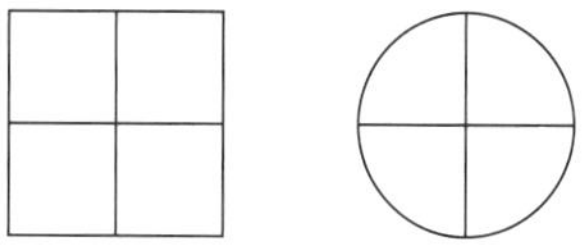

7. How many days are there in May?
8. How many halves are in a whole?
9. Write the number for two hundred and ninety-nine.
10. How many legs might a centipede have?
11. What is the fourth month of the year?
12.

+	3	6	9	12	15	18	21
3							

Measurement Centimetres

Draw lines of these lengths using the 1-cm dot paper.

a. 4 cm
b. 6 cm
c. 8 cm
d. 10 cm
e. 7 cm

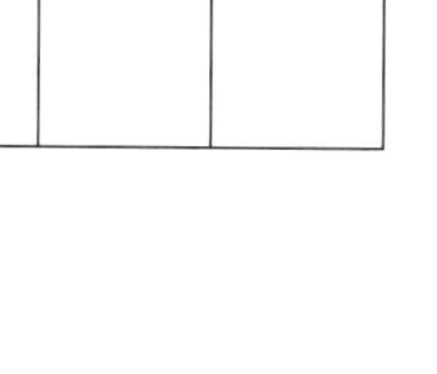

Number and Algebra

SET 1 Basic

		−2
1	3	
2	4	
3	5	
4	6	
5	7	
6	8	

7 9 + 2 = ☐

8 7 + 2 = ☐

9 9 − 2 = ☐

10 If Ashley bought two toy cars at one shop and three at another, how many did he buy altogether? ☐

SET 2 Addition to 99

Complete each addition.

1

	Tens	Ones
	5	4
+	2	3

2

	Tens	Ones
	6	5
+	2	4

3

	Tens	Ones
	5	3
+	2	4

4

	Tens	Ones
	6	3
+	2	3

5

	Tens	Ones
	4	8
+	3	1

6

	Tens	Ones
	7	7
+	2	2

7

	Tens	Ones
	4	4
+	2	4

8

	Tens	Ones
	3	4
+	2	5

Space Triangles

Join the dots to make triangles.

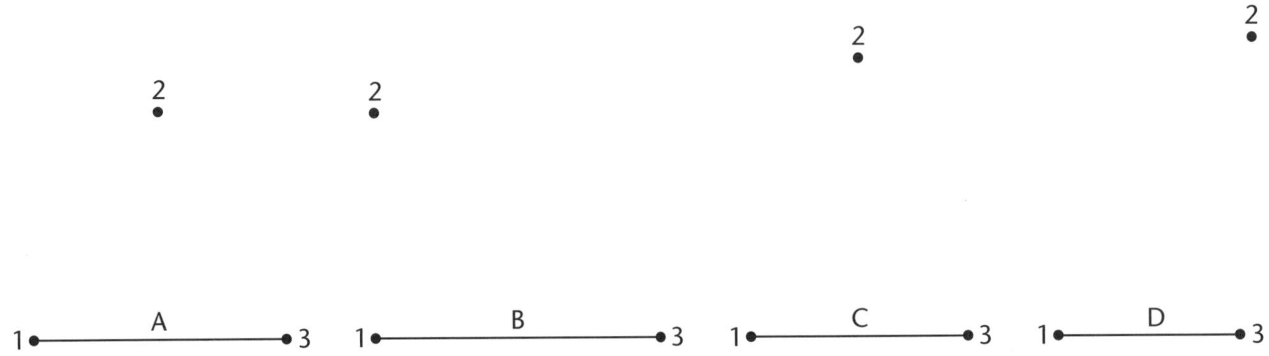

Number and Algebra

SET 3 Multiplication/groups of

Complete the number sentences.

1

3 groups of 4 = ☐

2

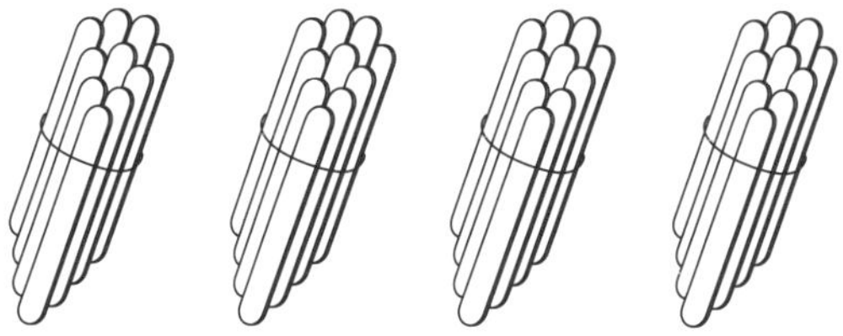

4 groups of 10 = ☐

3

5 groups of 3 = ☐

4

4 groups of 5 = ☐

SET 4 Extension

1 18 – 9 = ☐

2 10 rows of 4 equals ____.

3 How many legs do 3 birds have?

4 2 × 3

5 18, 20, 22, ____, ____, ____

6 27 = ☐ tens ☐ ones

7 What number comes after 99?

8 If  = 3 how many does  equal?

9 Draw four rows of 7 circles.

☐

10 How many sides have two squares?

11 What month is Christmas in?

12 21 – 4

13 Two less than 71

Measurement Duration

Use hours and minutes to describe the duration of these events.

	Start	Finish	Hours
1			
2			

	Start	Finish	Minutes
3			
4			

Number and Algebra

SET 1 Basic

1 3 + 2 = ☐

2 10 + 2 = ☐

3 6 + 1 = ☐

4 7 − 2 = ☐

5 3 + 3 = ☐

6 10 − 2 = ☐

7 4 + 4 = ☐

8 5 − 1 = ☐

9 9 − 4 = ☐

10 Harry had two dogs and four cats as pets. How many pets did he have altogether? ☐

SET 2 Making numbers to 100

Complete the number expanders for the modelled materials.

1

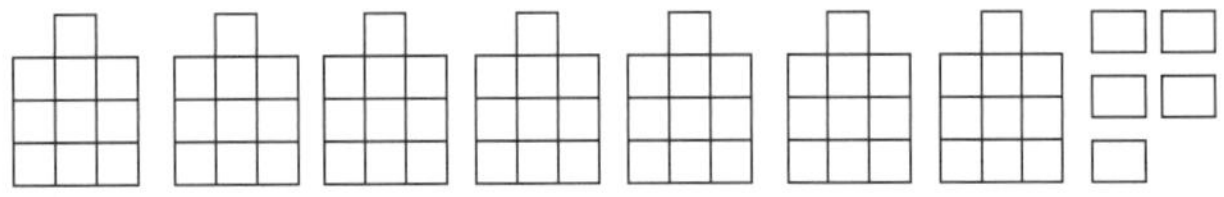

☐ tens ☐ ones

2

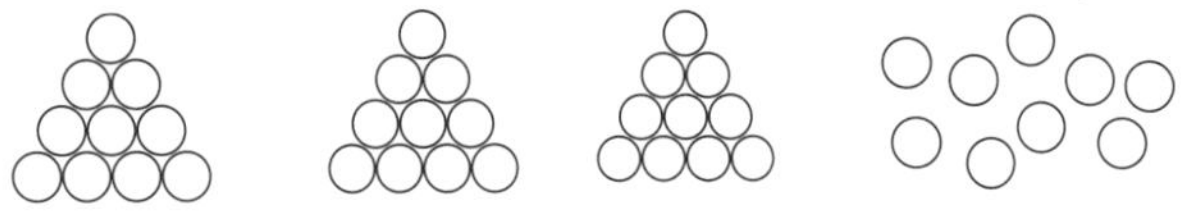

☐ tens ☐ ones

Draw strokes to model the numbers.

	Number	Hundreds	Tens	Ones
3	324	III		
4	681			
5	653			
6	702			
7	993			

Number and Algebra Addition jump strategy

Use the jump strategy to solve the problems.

1 Courtney had 15 cards and her dad gave her another 15 cards.

15 + 15 = ☐

15 16 17 18 19 20 21 22 23 24 25 26 27 28 29 30 31 32 33 34 35 36 37 38 39 40

2 How many children were on the bus if there were 32 girls and 19 boys?

32 + 19 = ☐

30 31 32 33 34 35 36 37 38 39 40 41 42 43 44 45 46 47 48 49 50 51 52 53 54

Number and Algebra

SET 3 Rounding and estimating

Round off these numbers to the nearest 10.

1 7 = ☐

2 13 = ☐

3 17 = ☐

4 26 = ☐

5 24 = ☐

6 36 = ☐

7 38 = ☐

8 25 = ☐

9 93 = ☐

10 45 = ☐

SET 4 Extension

1 3 tens + 6 ones = ☐

2 10 + 11

3 Which is larger: 13 + 2 or 15 – 1?

4 How many months are in a season?

5 22 – 7

6 6 + ☐ = 30

7 30c + 50c =

8 ☐ – 3 = 20

9 2, 4, 6, ____, 10, ____, 14

10 If 1 bag has 12 apples, how many are in 2 bags?

11 What date is Boxing Day?

12 4 less than 17

13 19 + 2

14 How many days in 2 weeks?

Space 2D faces

Colour the faces on the objects. Use the guide below.

Rectangles = yellow

Triangles = pink

Hexagons = blue

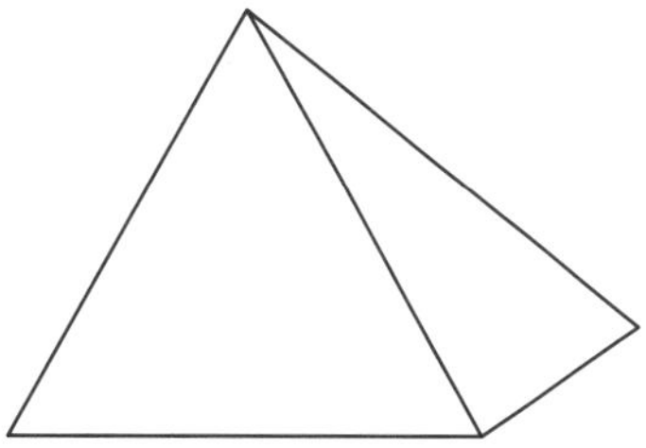

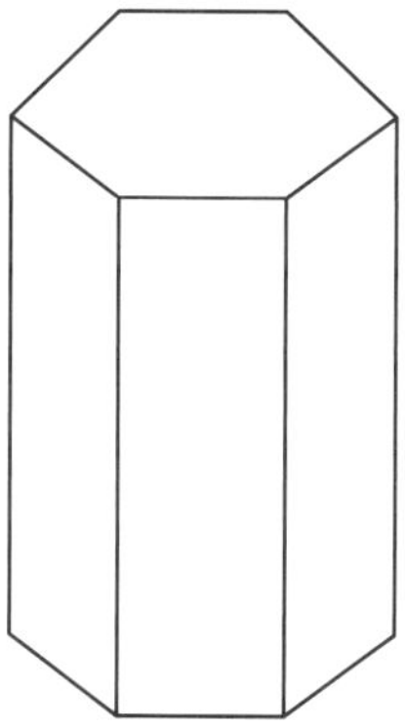

Number and Algebra

SET 1 Basic

	+ 3
1 2	
2 3	
3 4	
4 5	
5 6	
6 7	

7 10 + 5 = ☐

8 15 − 5 = ☐

9 6 + 4 = ☐

10

On Monday, Jenna swam 4 laps in the morning and 10 laps at night. How many laps did she swim altogether?

☐

SET 2 Multiplication/arrays

Write and solve a number sentence to describe the arrays.

1

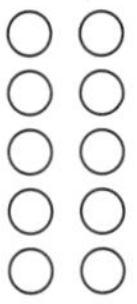

5 × ☐ = ☐

2

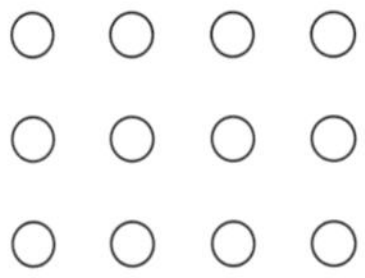

3 × ☐ = ☐

3

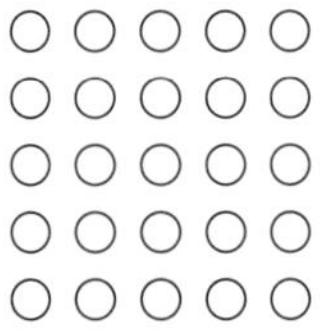

5 × ☐ = ☐

4

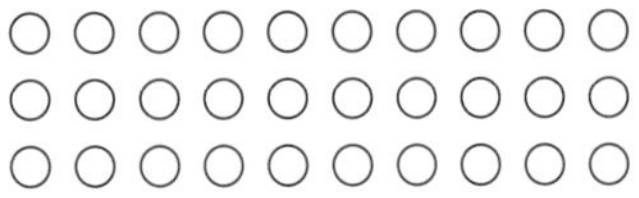

☐ × ☐ = ☐

Space Kites and rhombuses

Draw an example of each shape.

Square	Rhombus	Triangle	Kite

Number and Algebra

SET 3 Connecting problems to number sentences

Connect each piece of information in the problem to the number sentence, then find the answer.

1 There were 14 birds in a tree and 12 on the ground. How many were there altogether? □ + □ = □

2 Mia bought a soccer ball for $28 and boots for $49. How much did Mia spend? □ + □ = □

3 There were 31 students in class 2K and 27 in class 2P. How many students were there in total? □ + □ = □

SET 4 Extension

1 52, 62, ________, 82, ________, 102

2 How many quarters in a whole?

3 25 − □ = 10

4 What fraction of this shape has been shaded?

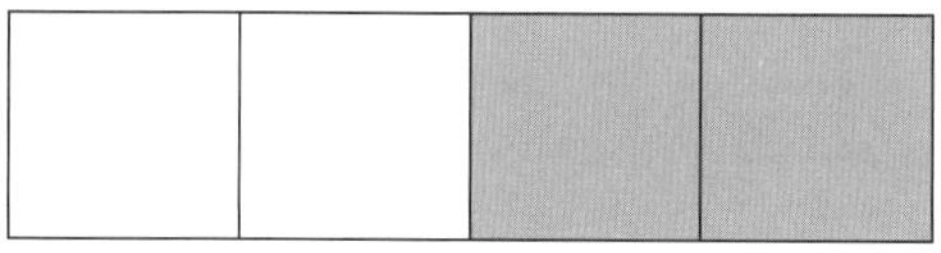

5 How many sides do 4 triangles have?

6 27 + □ = 33

7 Jayne had $1.20 but spent $0.70. How much does she have left?

8 Twelve less than 42

9 34, 44, ________, 64, ________, 84

10 36 + 54

11 How many centimetres in 1 m?

12 24 = □ dozen

13 Put these numbers in order from smallest to largest.
23, 83, 43, 13, 10, 22

Statistics Graphs

Favourite Foods of Class 2R

Colour the blocks to represent the foods. The first one has been done for you.

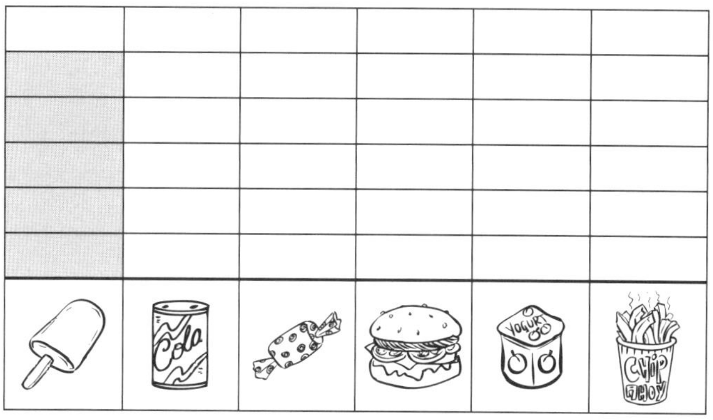

Number and Algebra

SET 1 Basic

1 [10] + [6] = [ ]

2 [10] + [2] = [ ]

3 [3] + [3] = [ ]

4 [4] + [2] = [ ]

5 [10] − [4] = [ ]

6 [9] + [1] = [ ]

7 [8] − [3] = [ ]

8 [10] + [5] = [ ]

9 [6] + [6] = [ ]

10 If there were 20 cows and 5 were sold, how many are left? [ ]

SET 2 Sharing

Use the popsticks to answer the sharing questions.

1 Twelve popsticks shared between [2] = [ ] popsticks

2 Twelve popsticks shared among [3] = [ ] popsticks

3 Twelve popsticks shared among [4] = [ ] popsticks

4 Twelve popsticks shared among [6] = [ ] popsticks

5 Twelve popsticks shared among [12] = [ ] popsticks

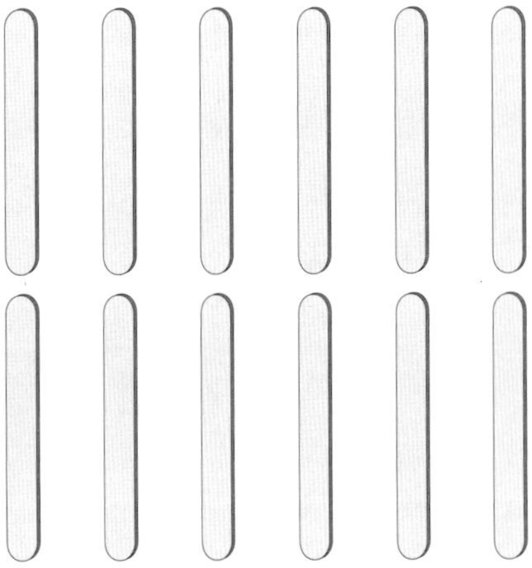

Number and Algebra Doubling and halving

Double the number in each group.

1

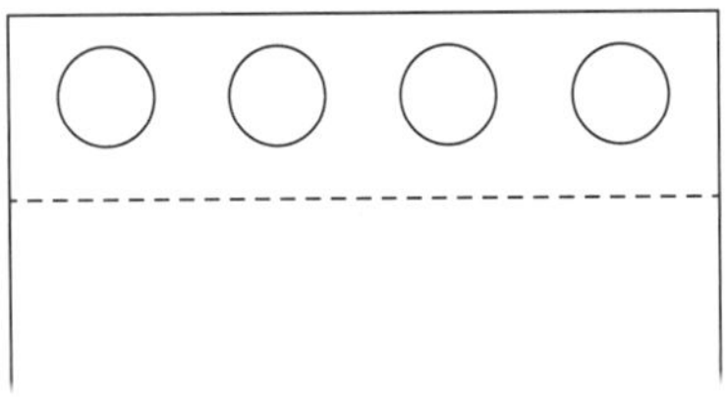

Double 4 = [ ]

2

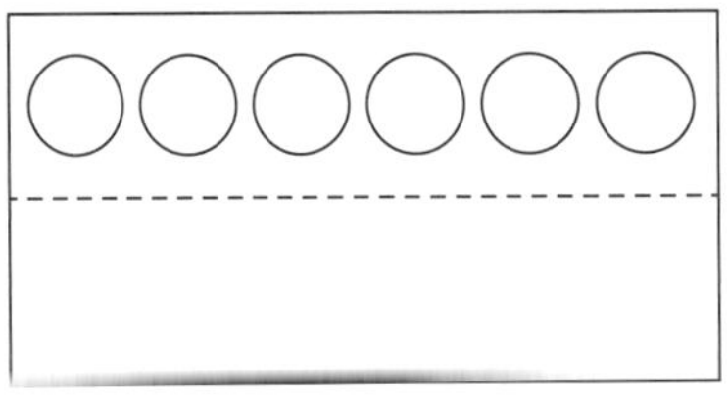

Double 6 = [ ]

3

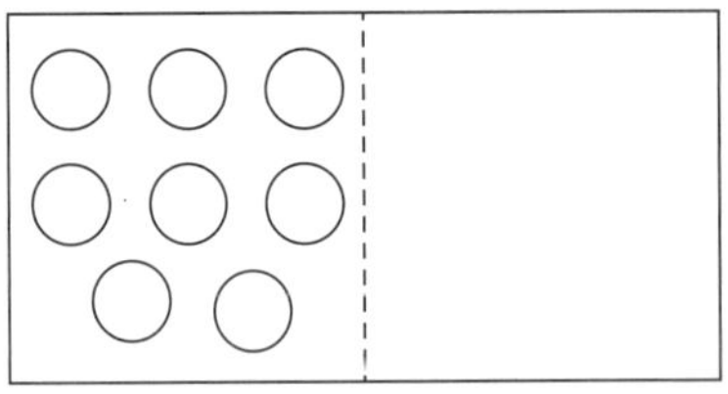

Double 8 = [ ]

Cross off half of the items in each group.

4

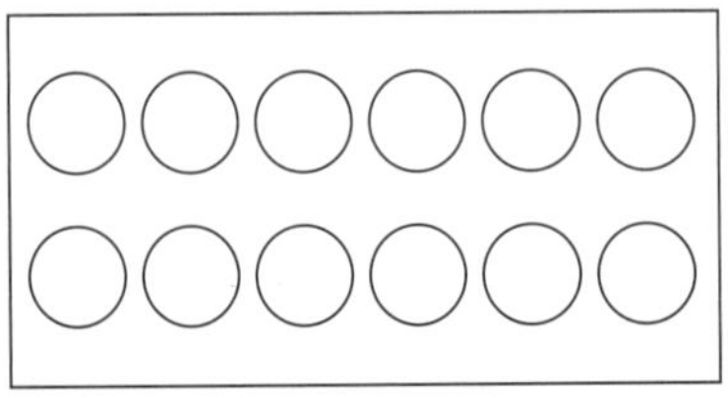

Half of 12 = [ ]

5

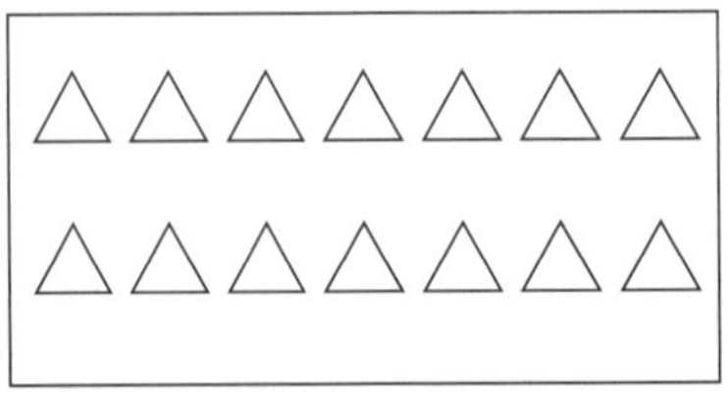

Half of 14 = [ ]

6

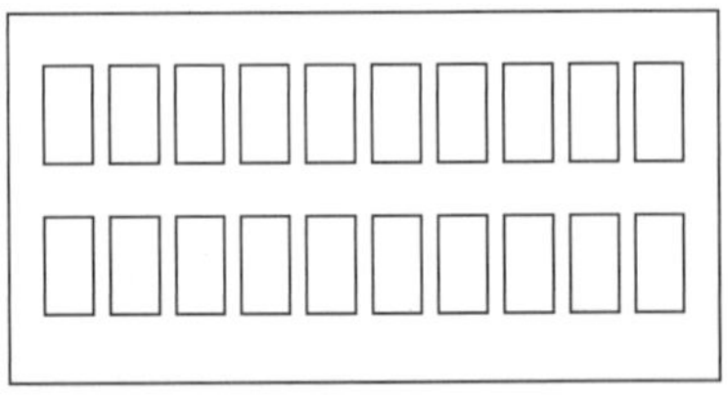

Half of 20 = [ ]

Number and Algebra

SET 3 Problems

Solve the problems.

1 Tarun and Luke each bought a packet of lollies. If there were 14 lollies in each packet, how many did they have altogether?

2 Linh and Aria wanted to go to a concert. Together they had $70. If the tickets were $35 each, did they have enough money to go?

3 Joe made 3 rows of counters. If there were 5 counters in each row, how many counters did he use?

4 There were 15 birds sitting on a branch. When they heard a loud bang, 8 of them flew away. How many birds were still on the branch?

SET 4 Extension

1 What is the coldest season?

2 $\frac{1}{2}$ of 12

3 45 + 12

4 Cents in $1.10

5 Fifteen less than 45

6 How many 20c coins in $1.20?

7 How many sides does a pentagon have?

8 7 tens 6 ones =

9 14 + 16

10 40, 44, 48, ____, ____

11 How many days in March?

12 99 – 44

13 How many legs on 2 horses?

14 45 + 43

Measurement Litres

Find containers that are measured in litres and list them below. The first one has been done for you.

Container	Litres
Milk carton	1

Number and Algebra

SET 1 Basic

		+4
1	2	
2	4	
3	6	
4	10	
5	12	
6	14	

7. 7 + 6 = ☐

8. 4 − 3 = ☐

9. 10 − 6 = ☐

10. Taryn planted 5 trees at school and 10 at the local park. How many did she plant altogether?

☐

SET 2 Place value

Use the number expander to record the number displayed on the abacus.

1
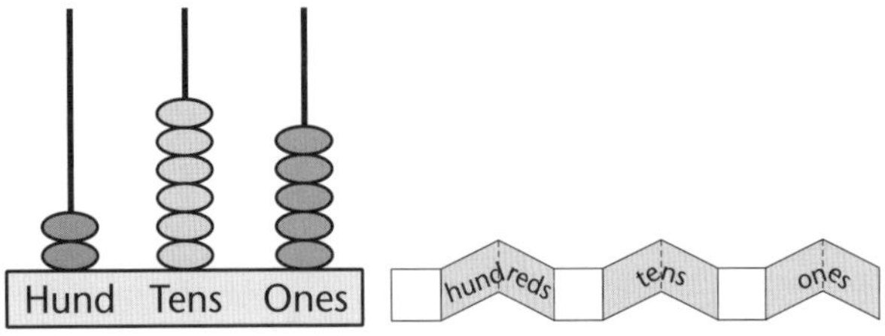

2
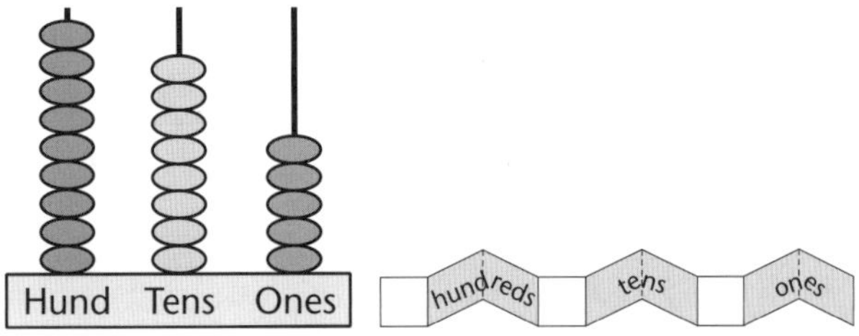

3
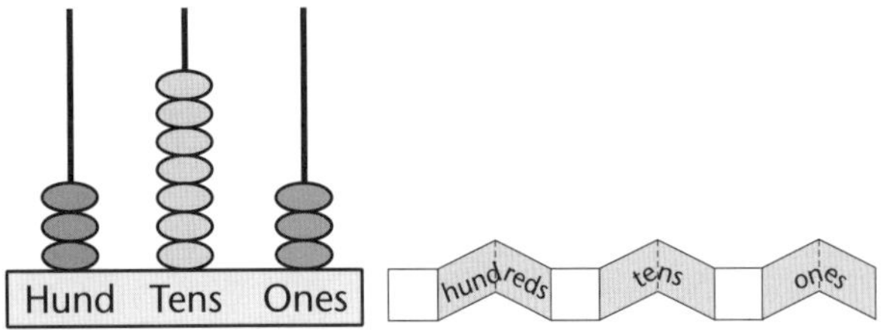

4
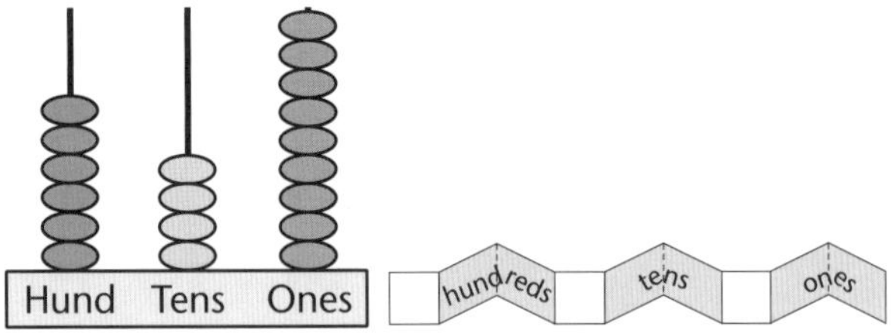

Number and Algebra Subtraction jump strategy

Use the jump strategy to solve the problems.

1. Purvi had 33 cards and she gave her dad 9 cards. How many cards did Purvi have left?

   33 − 9 = ☐

   15 16 17 18 19 20 21 22 23 24 25 26 27 28 29 30 31 32 33 34 35 36 37 38 39 40

2. How many children were on the bus if there were 46 and then 15 children got off?

   46 − 15 = ☐

   30 31 32 33 34 35 36 37 38 39 40 41 42 43 44 45 46 47 48 49 50 51 52 53 54

Number and Algebra

SET 3 Doubles and near doubles

Answer the doubles and near doubles questions.

1 9 + 9 = ☐

2 9 + 10 = ☐

3 12 + 12 = ☐

4 12 + 14 = ☐

5 20 + 20 = ☐

6 20 + 21 = ☐

7 35 + 35 = ☐

8 35 + 37 = ☐

9 40 + 40 = ☐

10 40 + 45 = ☐

SET 4 Extension

1 27 + 4

2 32 – ☐ = 23

3 8 tens plus 3 ones

4 How many sides on a triangle?

5 How many sides on 7 triangles?

6 How many hours in one day?

7 Which month comes before October?

8 60 – 12

9 Colour $\frac{3}{4}$ of these shapes. ☐ ☐ ☐ ☐

10 If this month is January, what month will it be in 3 months time?

11 Change from $2 if I spent 50c

12 10 + ☐ = 56

13 54 + 20

14 In what position did the fish come?

1st 2nd ☐ 4th

Space Regular and irregular hexagons

Join the dots to form the hexagons.

1

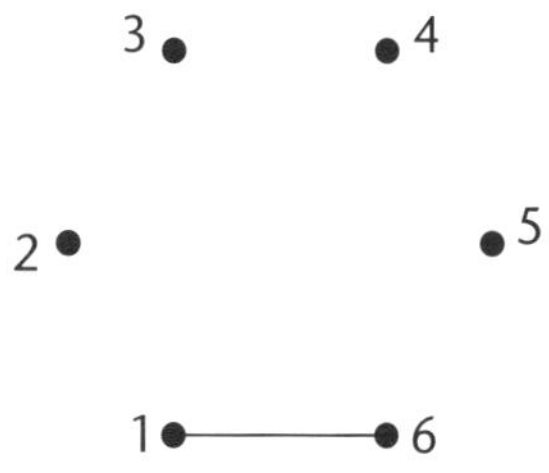

2

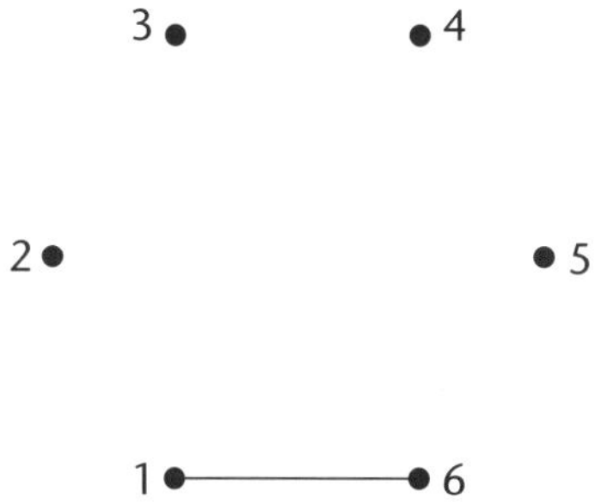

3

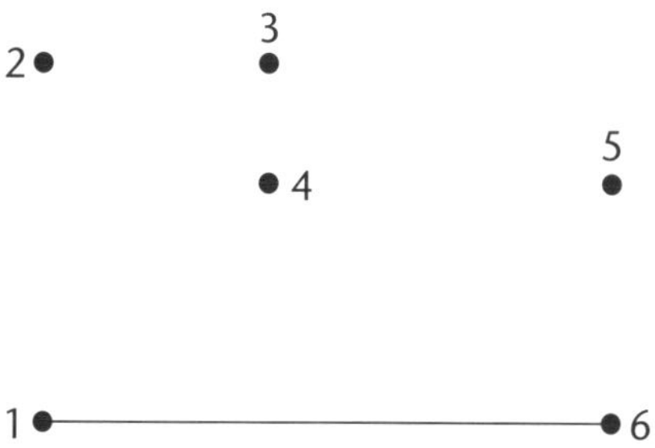

UNIT 14

Number and Algebra

SET 1 Basic

1 4 + 4 = ☐

2 10 − 6 = ☐

3 8 + 3 = ☐

4 10 + 2 = ☐

5 5 + 1 = ☐

6 6 − 2 = ☐

7 Number after 7 = ☐

8 Number before 7 = ☐

9 Double 4 = ☐

10

Bree and Grant each bought 6 new games for their computer. How many do they have altogether?

☐

SET 2 Subtraction to 99

Subtract 2-digit numbers.

1

	Tens	Ones
	8	0
−	6	0

2

	Tens	Ones
	7	8
−	3	5

3

	Tens	Ones
	5	0
−	4	0

4

	Tens	Ones
	3	9
−	2	7

5

	Tens	Ones
	9	7
−	7	0

6

	Tens	Ones
	4	8
−	4	4

7

	Tens	Ones
	8	7
−	3	0

8

	Tens	Ones
	5	9
−	3	4

Space Sketching position

Petri built a model out of the blocks. Put the letters on the blocks to discover the secret message.

1 Put a **C** on the top left block.

2 Put an **N** on the middle block.

3 Put a **D** on the bottom left block.

4 Put a **T** on the top right block.

5 Put a **G** on the bottom right block.

6 Put another **D** to the right of **N**.

7 What is the message? ______________________

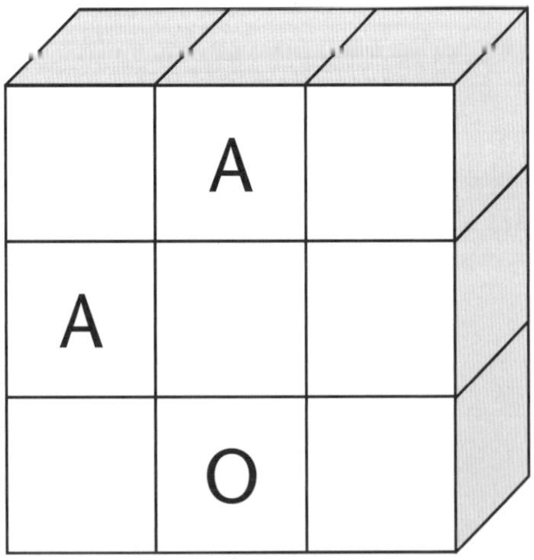

Number and Algebra

SET 3 Eighths

Shade the fractions of each shape.

1 $\frac{1}{8}$

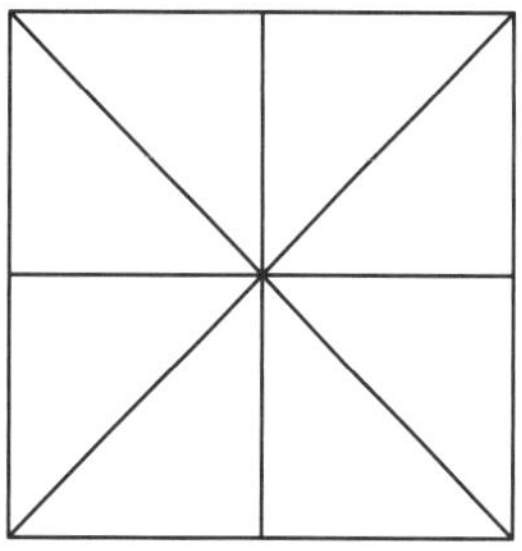

2 $\frac{1}{4}$

3 $\frac{3}{8}$

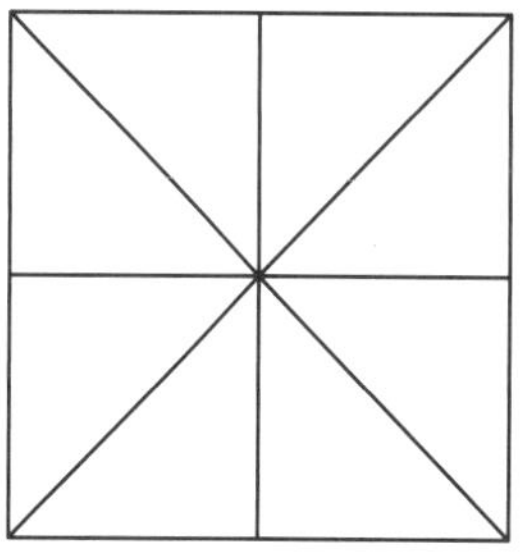

4 Which is the larger fraction $\frac{1}{8}$, $\frac{1}{4}$ or $\frac{3}{8}$? ____________________

SET 4 Extension

1

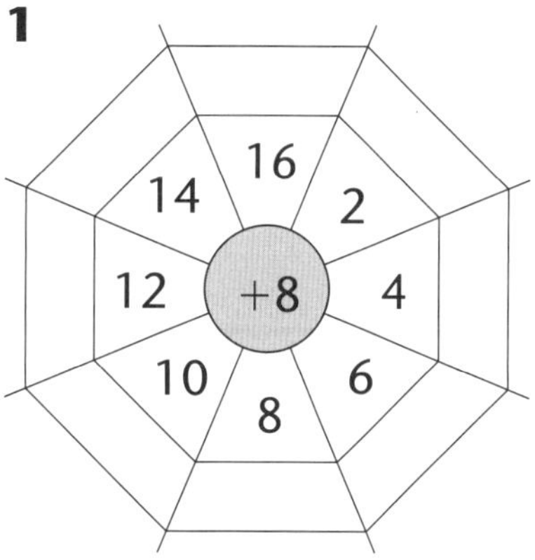

2 37 − 11 =

3 Is 21 odd or even?

4 Write three hundred and two.

5 ☐ + 13 = 80

6 Double 12.

7 5 × 4

8 39 + 21

9 How many in 3 dozen?

10 \$5 − \$1.50

11 98 = ☐ tens ☐ ones

12 Colour the heart 4 from the left.

Measurement Time

Draw a line to match the clock faces to the times.

6 o'clock

Half past 3

Half past 10

Quarter past 1

Quarter to 7

Quarter past 10

Number and Algebra

SET 1 Basic

		+ 5
1	2	
2	4	
3	5	
4	7	
5	10	
6	15	

7 6 + 3 = ☐

8 1 + 5 = ☐

9 12 − 5 = ☐

10

Luke owned 3 cars and his sister Lauren owned 3. How many cars did they own altogether?

☐

SET 2 Multiplication facts (2s)

☆☆★★☆☆★★☆☆★★☆☆★★☆☆★★

Answer the multiplications. The stars may help you.

1 2 × 2 = ☐

2 3 × 2 = ☐

3 1 × 2 = ☐

4 4 × 2 = ☐

5 5 × 2 = ☐

6 7 × 2 = ☐

7 6 × 2 = ☐

8 8 × 2 = ☐

Space Classifying 2D shapes

Join the dots to form a 2D shape then name it and record the number of sides and corners on each shape.

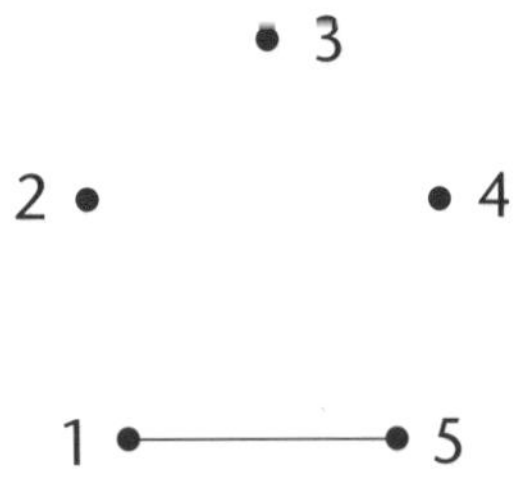

Name

Sides ☐ Corners ☐

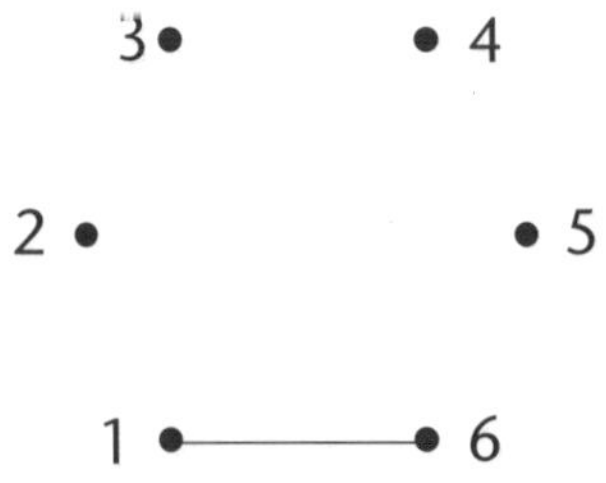

Name

Sides ☐ Corners ☐

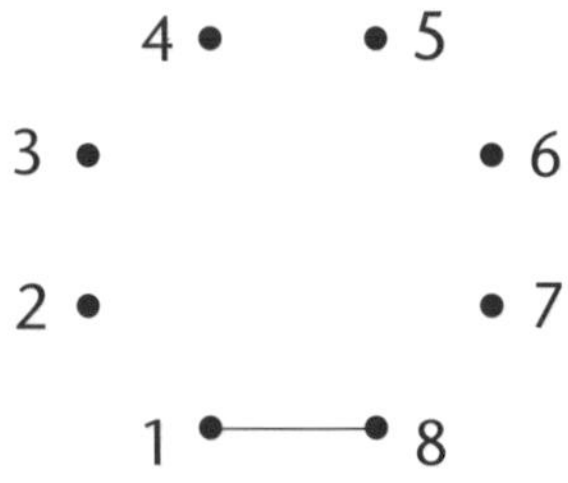

Name

Sides ☐ Corners ☐

Number and Algebra

SET 3 Number bond combinations to 20

Complete the number bonds.

1 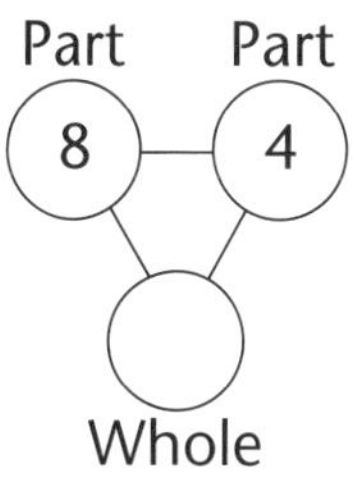

2 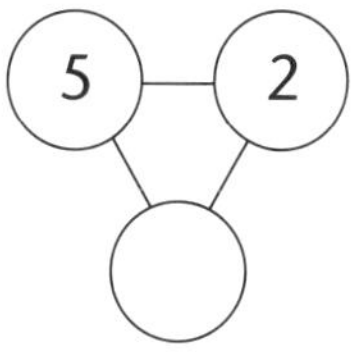

3 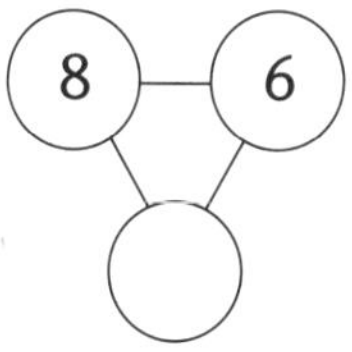

4 Use number bonds to find the whole.

4 + 3	

5 + 2	

6 + 1	

SET 4 Extension

1 How many sides in 2 squares and 1 rectangle?

2 22 + ☐ = 100

3 If Janice fell asleep at 9:00 pm and only slept for 3 hours, at what time did she wake up?

4 5 + 12 + 6

5 Which season is the hottest?

6 44 divided by 2

7 15 lollies shared between 3 people = ☐ lollies each

8 78 = ☐ tens ☐ ones

9 50 – 12

10 2 + 4 + 6

11 9 groups of 2

12 $\frac{50}{100} - \frac{1}{\quad}$

13 How many hours in 2 days?

14 Is 50 an odd or an even number?

Measurement Mass

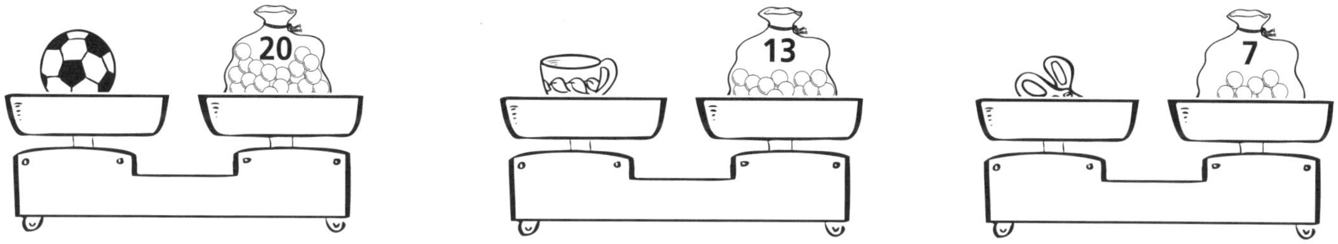

Each item's mass was measured with a bag of marbles. What is the difference in mass between:

1 the ball and the cup? ☐ marbles

2 the ball and the scissors? ☐ marbles

3 the cup and the scissors? ☐ marbles

4 Would the ball be balanced if the cup and the scissors were placed in the other pan? ☐

Number and Algebra

SET 1 Basic

1 5 + 3 = ☐

2 4 − 2 = ☐

3 6 + 4 = ☐

4 8 + 5 = ☐

5 10 + 5 = ☐

6 15 − 5 = ☐

7 One more than 12 = ☐

8 Double 3 = ☐

9 One less than 6 = ☐

10 Jack's dad had 13 computers at his office. If 3 broke down, how many still work? ☐

SET 2 Addition jump strategy

Add the following numbers by adding the tens first, then adding the ones. The first one is done for you.

1	34 + 13	becomes	34	+	10	+	3	=	47
2	36 + 22	becomes		+		+		=	
3	24 + 23	becomes		+		+		=	
4	43 + 24	becomes		+		+		=	
5	52 + 24	becomes		+		+		=	

34 + 23 becomes 34 + 20 + 3 = 57

Statistics Picture graphs

1 Which hair colour was the most common?

____

2 Which hair colour was the least common?

____

3 How many more students had blond hair than black? ______________________________

4 How many more students had brown hair than fair? ______________________________

Class 2T Hair Colours

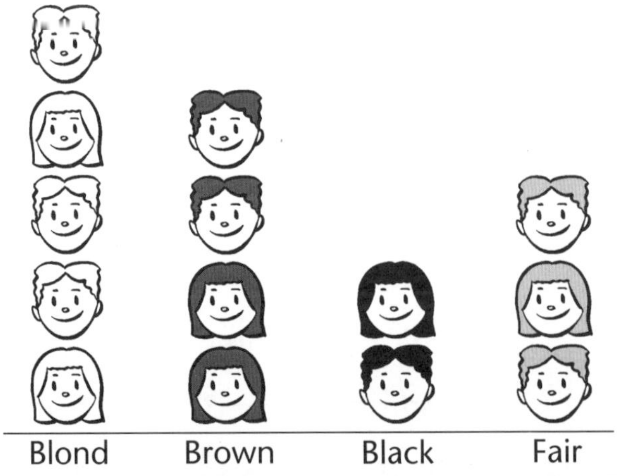

Number and Algebra

SET 3 Sharing using groups

Answer the divisions. The array of faces may help you.

1  16 divided by 4 =	**2**  12 divided by 3 =	**3**  20 divided by 4 =

SET 4 Extension

1

+	1	2	3	4	10	11
9						

2 5 less than 80

3 How many days in a year?

4 How many legs are there on 3 cows?

5 10 + 3 + 13

6 How many months in 2 seasons?

7 ☐ − 20 = 120

8 20 divided by 2

9 25 + 25

10 How many sides do two triangles and three squares have altogether?

11 19 − 7

12 3 × 3

13 Write three hundred and thirty-five dollars.

Measurement Centimetres

Measure the length of each arrow in centimetres.

1 

2 

3 

4 

5 

6 

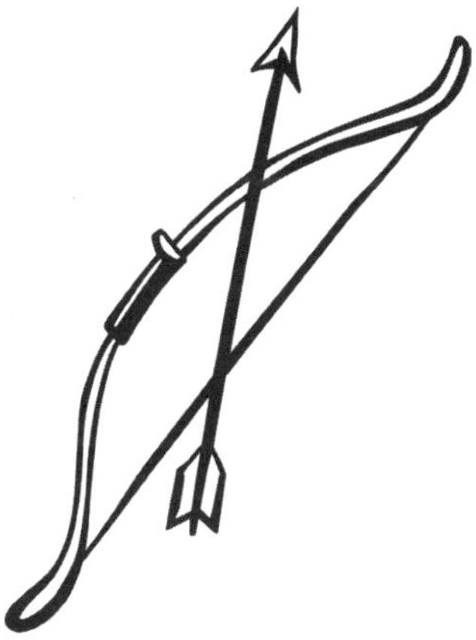

Number and Algebra

SET 1 Basic

		−3
1	3	
2	5	
3	6	
4	8	
5	10	
6	15	

7 7 + 3 = ☐

8 10 + 3 = ☐

9 18 − 3 = ☐

10 If 20 people tried out for a play but there were only five parts, how many will miss out? ☐

SET 2 Multiplication–turnarounds

Write two multiplication facts for each array.

1

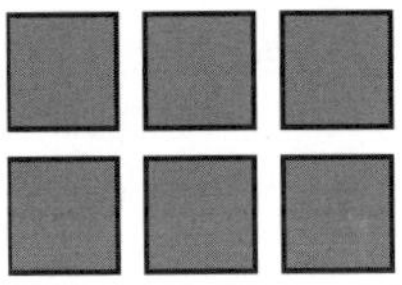

☐ × ☐ = ☐

☐ × ☐ = ☐

2

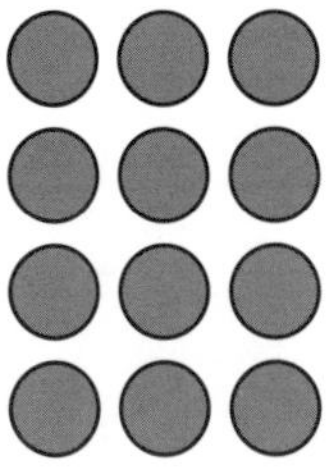

☐ × ☐ = ☐

☐ × ☐ = ☐

3

☐ × ☐ = ☐

☐ × ☐ = ☐

Number and Algebra Locating numbers on a number line

Which number in the box describes the position of the X on each number line?

1 ☐

2	12	10

2 ☐

70	90	50

Number and Algebra

SET 3 Extending addition facts

Complete the patterns to extend these addition facts.

1

	12	13	14	15	16	17	18
+	10	10	10	10	10	10	10

2

	14	24	34	44	54	64	74
+	6	6	6	6	6	6	6

3

	12	22	32	42	52	62	72
+	5	5	5	5	5	5	5

4

	24	34	44	54	64	74	84
+	7	7	7	7	7	7	7

SET 4 Extension

1 9 + 3 + 2

2 48 hours = ☐ days

3 15c + 20c + \$1

4 2 × 7

5 Total of 15 and 7

6 300, 400, ☐, 600, ☐

7 Subtract 20c from \$2.

8 \$5 – \$2.50

9 2 tens plus 3 ones

10 46 to the nearest 10

11 8, ☐, 4, 2

12 What is the sum of three and ten?

13 Colour the heart that is two in from the left red, and the heart four in from the right blue.

Space Parallel lines and sides

1 Colour the shapes that have parallel sides.

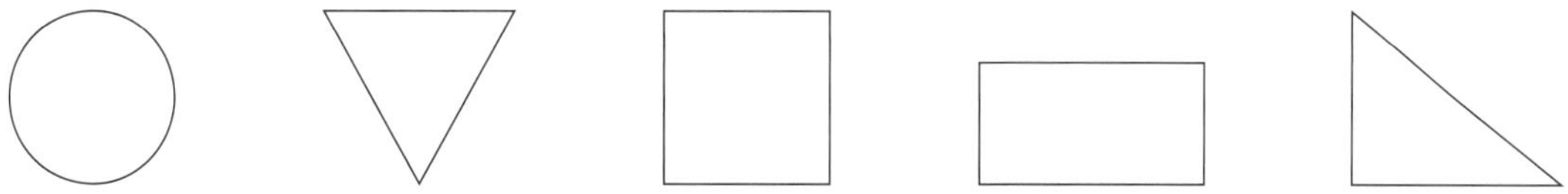

2 Draw a line parallel to the one in each box.

UNIT 18

Number and Algebra

SET 1 Basic

1 7 − 1 = ☐

2 3 + 3 = ☐

3 4 + 3 = ☐

4 10 − 5 = ☐

5 8 − 3 = ☐

6 6 + 4 = ☐

7 One more than 11 = ☐

8 Is 18 an odd number? ☐

9 2 × 4 = ☐

10

If there are 16 pages in a book and Phil has read 3, how many are left?

☐

SET 2 Number patterns

Complete the counting patterns below by following the rules.

1 Count by 2s.

0	2	4		8		12	14	16	

2 Count by 5s.

0	5	10		20	25		30	35	

3 Count by 10s.

0	10	20		40		60	70		90

4 Count by 3s.

0	3	6		12	15	18		24	

Space 2D shapes

Draw a line to match the description and name.

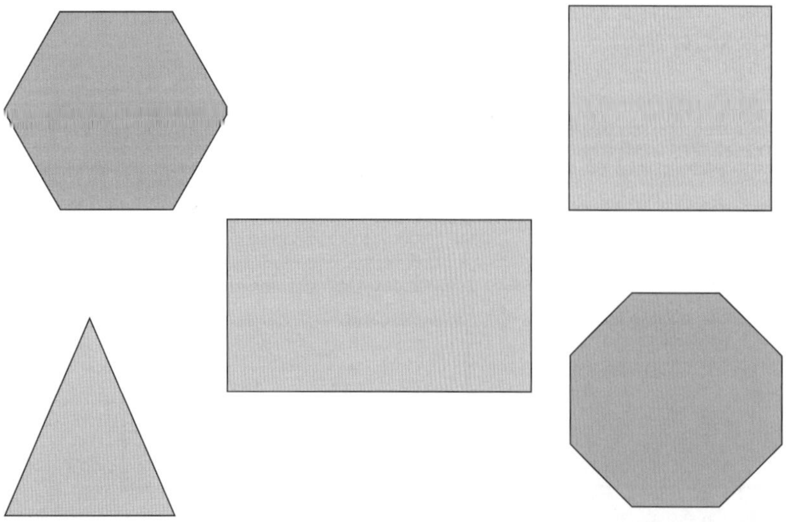

Rectangle
4 sides and 4 corners

Hexagon
6 sides and 6 corners

Square
4 sides and 4 corners

Triangle
3 sides and 3 corners

Octagon
8 sides and 8 corners

Number and Algebra

SET 3 Bridging to 10

Use the strategy outlined or any other strategy to answer the questions.

1	8 + 3 = ☐	**6**	9 + 6 = ☐
2	3 + 6 = ☐	**7**	7 + 4 = ☐
3	9 + 2 = ☐	**8**	8 + 6 = ☐
4	8 + 4 = ☐	**9**	6 + 8 = ☐
5	8 + 8 = ☐	**10**	5 + 7 = ☐

SET 4 Extension

1 Is 26 odd or even?

2 28 + ☐ = 35

3 7 less than 9

4 7 more than 9

5 6 + 3 + 2

6 22 + 26

7 Four thirty in digital time

8 How much are 3 envelopes at 50c each?

9 95 = ☐ tens ☐ ones

10 Total of 30 and 300

11 Which day follows Sunday?

12 Which is greater: 2 dozen or 21?

13 Round 42 to the nearest 10.

Statistics Data

There are 21 different coloured marbles in the bag that are red, blue and green.

Use trial and error to calculate your answer, then colour the marbles so that:

There are 8 blue marbles.

Red is less likely to be drawn out of the bag.

Green's likelihood is in between red and blue.

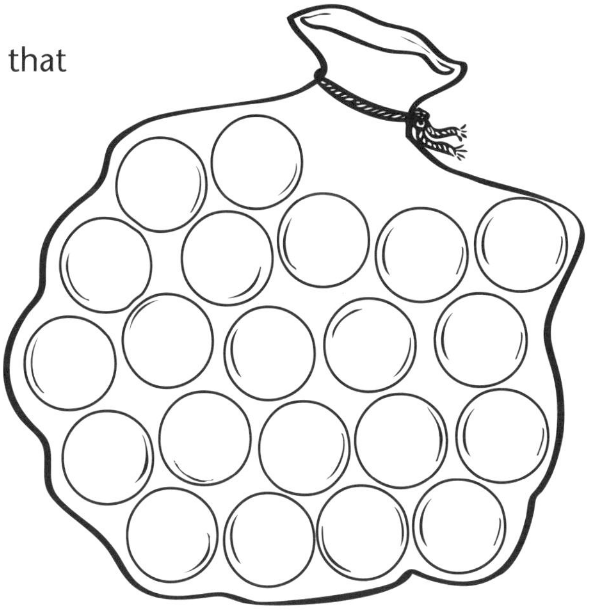

Number and Algebra

SET 1 Basic

	+6	
1	1	
2	2	
3	3	
4	4	
5	5	
6	10	

7 10 − 6 = ☐

8 6 + 6 = ☐

9 8 − 2 = ☐

10 What is half of 12? ☐

SET 2 Counting patterns

Complete the counting patterns below by following the rules.

1 Count by 5s.

45						

2 Count by 10s.

30						

3 Count by 2s.

12						

4 Count by 4s.

8						

5 Count by 5s.

95						

6 Count by 2s.

46						

7 Count by 10s.

110						

Measurement Calendar

September

S	M	T	W	T	F	S
30						1
2	3	4	5	6	7	8
9	10	11	12	13	14	15
16	17	18	19	20	21	22
23	24	25	26	27	28	29

October

S	M	T	W	T	F	S
	1	2	3	4	5	6
7	8	9	10	11	12	13
14	15	16	17	18	19	20
21	22	23	24	25	26	27
28	29	30	31			

November

S	M	T	W	T	F	S
				1	2	3
4	5	6	7	8	9	10
11	12	13	14	15	16	17
18	19	20	21	22	23	24
25	26	27	28	29	30	

December

S	M	T	W	T	F	S
30	31					1
2	3	4	5	6	7	8
9	10	11	12	13	14	15
16	17	18	19	20	21	22
23	24	25	26	27	28	29

What day of the week is:

1 October 19? ______

2 September 28? ______

3 December 25? ______

4 November 16? ______

5 9 days after November 13? ______

6 12 days after October 7? ______

Number and Algebra

SET 3 100s chart addition and subtraction

1	2	3	4	5	6	7	8	9	10
11	12	13	14	15	16	17	18	19	20
21	22	23	24	25	26	27	28	29	30
31	32	33	34	35	36	37	38	39	40
41	42	43	44	45	46	47	48	49	50
51	52	53	54	55	56	57	58	59	60
61	62	63	64	65	66	67	68	69	70
71	72	73	74	75	76	77	78	79	80
81	82	83	84	85	86	87	88	89	90
91	92	93	94	95	96	97	98	99	100

Solve the number sentences using the hundreds chart.

1. 10 + 13 = ☐
2. 21 + 14 = ☐
3. 31 + 27 = ☐
4. 54 − 21 = ☐
5. 81 − 36 = ☐

SET 4 Extension

1. What is the sum of 3 + 4 + 10?
2. Write the largest 3-digit number you can using 7, 9, 2.
3. 36 + 4 tens
4. $1.50 + $3
5. Write the smallest 3-digit number you can using 3, 4, 1.
6. ☐ tens + 3 ones = 73
7. $\frac{1}{2}$ of 6
8. 10, 20, 30, 40, 50
   Rule: ______________________
9. 25 + 75
10. ☐ + 3 = 100
11. Add 30 to 47.
12. What is the product of two and three?
13.

−	10	13	16	19	20	50
3						

Space Different orientations

Colour the shape that is the same size and shape as the shape in the box.

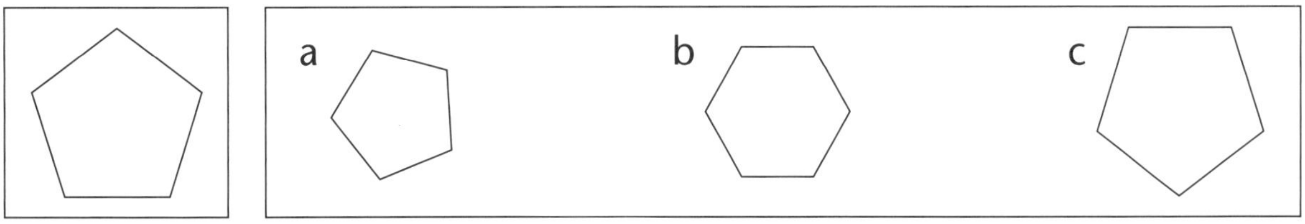

Number and Algebra

SET 1 Basic

1 4 − 4 = ☐

2 6 + 4 = ☐

3 5 − 2 = ☐

4 12 + 5 = ☐

5 15 − 5 = ☐

6 17 − 2 = ☐

7 5 less than 15 = ☐

8 One less than 13 = ☐

9 3 × 3 = ☐

10

Jasmine has $8.
If Danielle has $3 more, how much does Danielle have?

☐

SET 2 Repeated addition

Use the number lines to help you answer the questions.

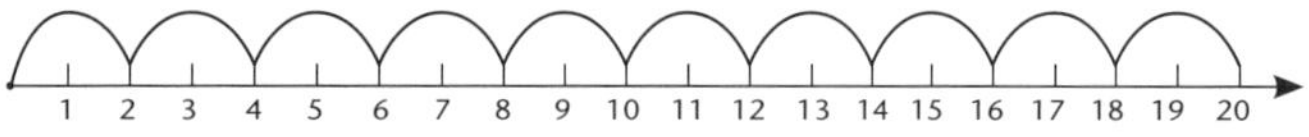

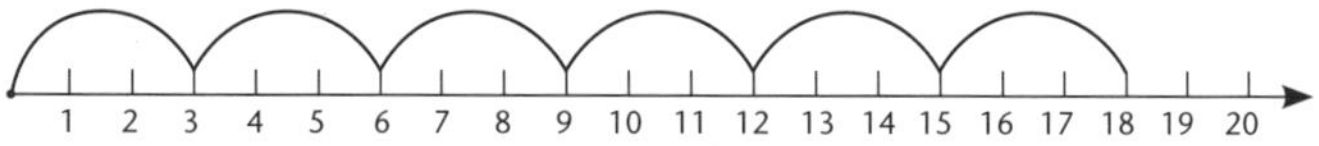

1 3 × 2 = ☐

2 5 × 2 = ☐

3 4 × 2 = ☐

4 6 × 2 = ☐

5 2 × 3 = ☐

6 4 × 3 = ☐

7 5 × 3 = ☐

Measurement Halves, quarters and eighths of a length

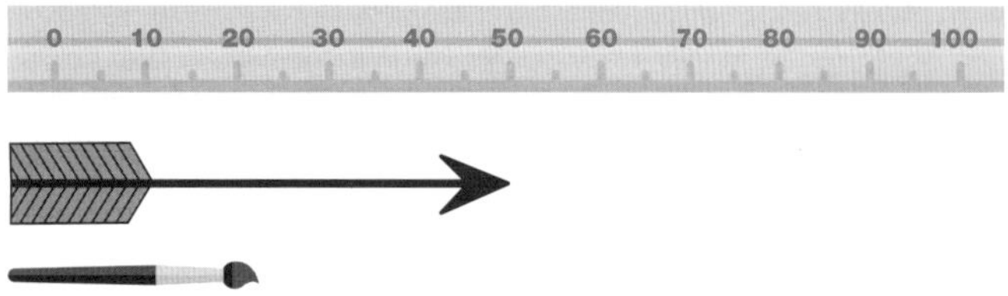

1 Colour the item that is one quarter the length of the line.

2 Circle the item that is one half the length of the line.

3 Tick the item that is one eighth the length of the line.

Number and Algebra

SET 3 Additive number patterns

Complete each number pattern, then write a rule for it.

1

2	4	6		10			16

Rule: ____________________

2

20	21	22			25		27

Rule: ____________________

3

40	35	30			15		5

Rule: ____________________

4

16	26	36		56			86

Rule: ____________________

Now make your own up and write its rule below.

5

Rule: ____________________

SET 4 Extension

1 7 + 8 + 7

2 Cost of a $10 cake and a $3 tart

3 How many days in 3 weeks?

4 16 ÷ 2

5 90, 100, 110, ☐, 130

Rule: ____________________

6 36 + 20

7 How many days in a school week?

8 How many months in 2 years?

9 50 – 5 – 5

10 Write the smallest number you can using 7, 6, 8.

11 Write the largest number you can using 7, 6, 8.

12 **a**

	Tens	Ones
	1	5
+	1	7

b

	Tens	Ones
	1	2
+	2	9

Space Quarter and half turns

Draw the turns:

a $\frac{1}{4}$ turn clockwise

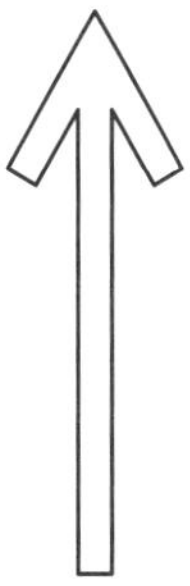

b $\frac{1}{4}$ turn anticlockwise

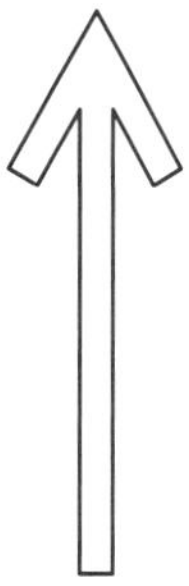

c $\frac{1}{2}$ turn clockwise

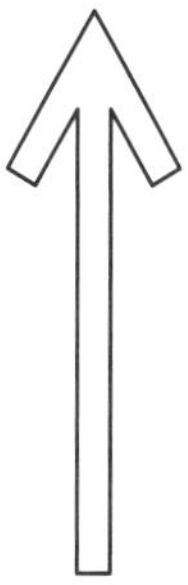

UNIT 21

Number and Algebra

SET 1 Basic

		− 4
1	5	
2	6	
3	10	
4	15	
5	16	
6	20	

7 12 − 2 = ☐

8 14 − 4 = ☐

9 10 + 4 = ☐

10 Jana had $15 but spent $4 on a book. How much does she have left? ☐

SET 2 Ordering numbers

Order the numbers from smallest to largest.

1	86	68	38	
2	22	32	26	
3	116	106	126	
4	333	393	383	
5	76	74	176	

What number is:

6 5 more than 72? ☐

7 5 less than 83? ☐

8 5 more than 22? ☐

9 5 less than 38? ☐

Number and Algebra Addition number bonds

Create 3 addition combinations that equal the number in the circle.

18: 10 + ☐ | 12 + ☐ | 15 + ☐

20: 16 + ☐ | 13 + ☐ | 2 + ☐

19: 14 + ☐ | 3 + ☐ | 17 + ☐

Number and Algebra

SET 3 Finding the difference

Find the difference between the two numbers.

1	10	and	4	=	**5**	14	and	6	=
2	10	and	6	=	**6**	20	and	5	=
3	11	and	3	=	**7**	16	and	4	=
4	12	and	2	=	**8**	19	and	7	=

SET 4 Extension

1

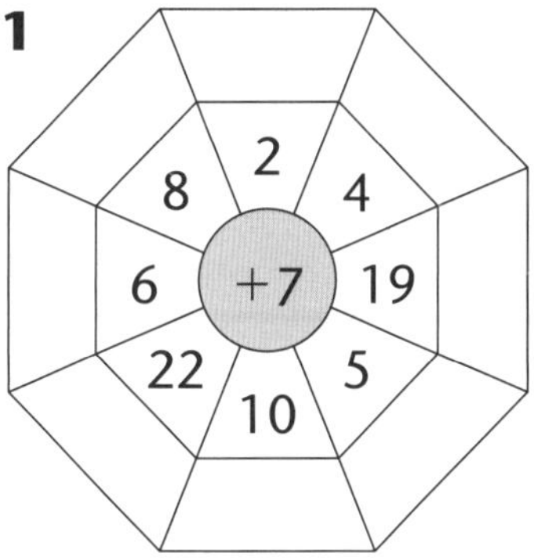

2 250c = $ ☐ . ☐ ☐

3 30 + 30 + 30

4 How many 10c pieces are in $1?

5 Is 72 odd or even?

6 13, 15, ____, 19, 21, ____

7 21 × 0

8 How many groups of 5 can be made from 20 marbles?

9 What is the possibility of snow today?

10 Make up 3 number sentences.

☐ + ☐
☐ + ☐ = 12
☐ − ☐

Statistics Graphs

Use the information on Class 2K's hair colours to colour the blocks and construct the graph.

Blond hair	**6**
Brown hair	**8**
Black hair	**5**
Red hair	**3**
Fair hair	**5**
Green hair	**0**

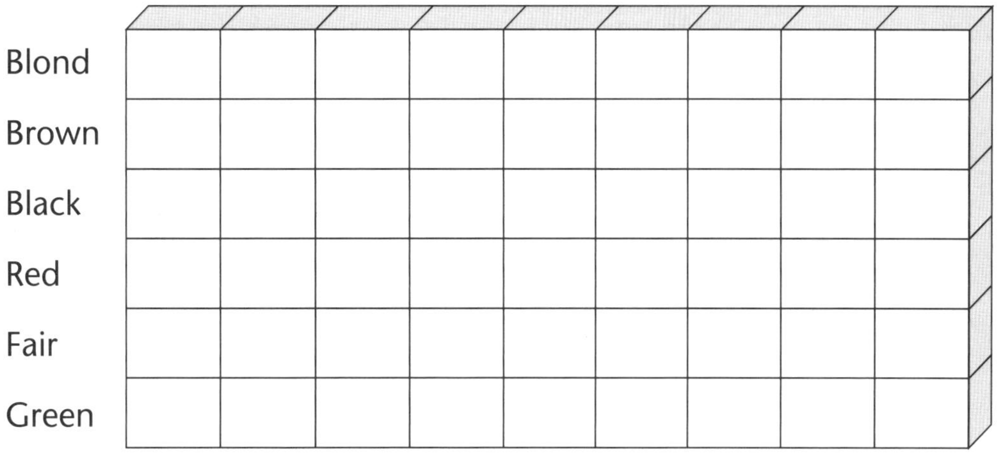

Number and Algebra

SET 1 Basic

1 10 − 5 = [ ]

2 3 + 2 = [ ]

3 5 − 1 = [ ]

4 8 + 3 = [ ]

5 14 + 5 = [ ]

6 6 − 4 = [ ]

7 Double 5 = [ ]

8 One more than 19 = [ ]

9 Is 10 an even number? [ ]

10

John had a pack of 20 stamps but used 3 to send a parcel. How many does he have left?

[ ]

SET 2 Rearranging numbers including zero

1
2 hundreds 6 tens 1 ones
[ ] [ ] tens [ ] ones
[ ] [ ] [ ] ones

2
3 hundreds 4 tens 3 ones
[ ] [ ] tens [ ] ones
[ ] [ ] [ ] ones

3
5 hundreds 0 tens 0 ones
[ ] [ ] tens [ ] ones
[ ] [ ] [ ] ones

4
7 hundreds 7 tens 9 ones
[ ] [ ] tens [ ] ones
[ ] [ ] [ ] ones

Space Nets of 3D objects

Write down everything you know about this object. Its net may help you.

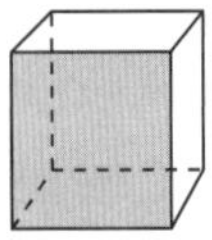

____

____

____

____

____

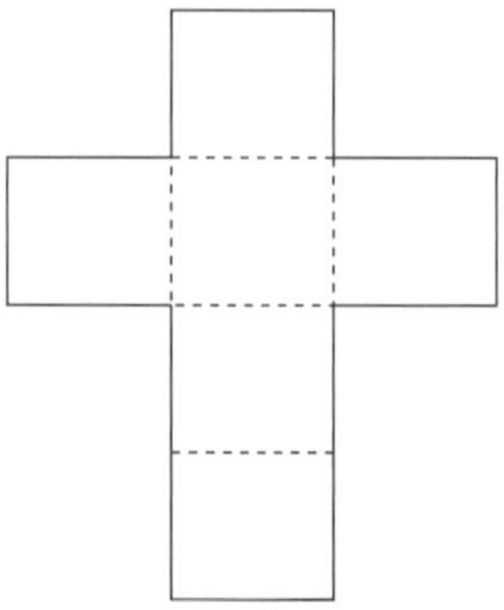

Number and Algebra

SET 3 Equal number sentences

Colour the correct labels to compare the number sentences.

1 8 + 7 = 20 − 5 True False

2 20 − 7 = 9 + 5 True False

3 22 + 13 = 30 + 5 True False

4 44 − 21 = 16 + 7 True False

5 60 + 7 = 30 − 4 True False

SET 4 Extension

1 ☐ + 7 = 25

2 9 + 4 + 6

3 Half of 40

4 $10 – 50c

5 Round 53 to the nearest 10.

6 How many legs do 4 cows have?

7 What is the value of 7 in 374?

8 ☐ × 10 = 20

9 How many 5c pieces in 30c?

10 How many halves in 2 wholes?

11 On the diagram below, label Jasmine in 3rd place, Kelly in 5th and Kristy in 7th.

Measurement Volume of objects

How many blocks make up the volume of these solid prisms?

1 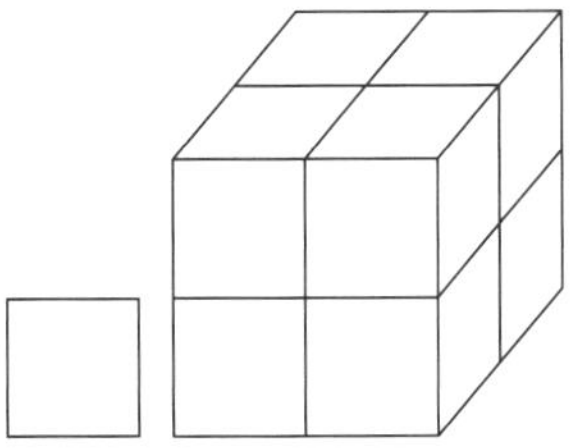

2 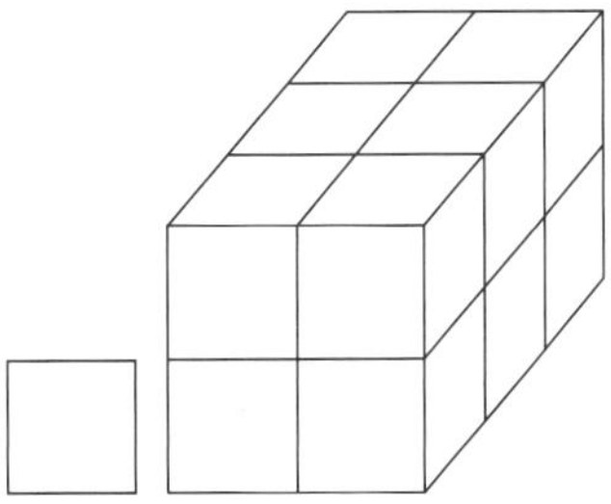

3 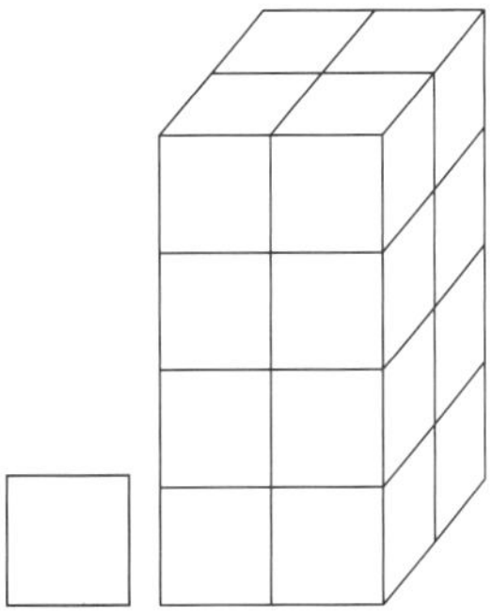

4 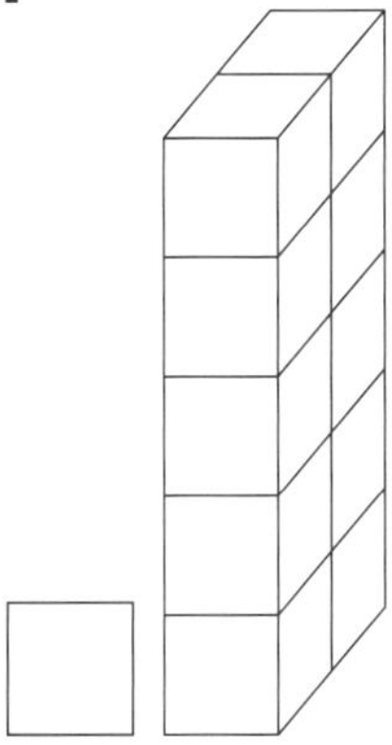

UNIT
23

Number and Algebra

SET 1 Basic

		+ 7
1	1	
2	2	
3	3	
4	5	
5	10	
6	13	

7 17 − 7 = ☐

8 7 + 3 = ☐

9 10 + 7 = ☐

10 Lauren had $15 but had to lend $10 to Kelly. How much does Lauren have left? ☐

SET 2 Subtraction jump strategy

Subtract the following numbers by taking away the tens first, and then the ones. The first one has been done for you.

1	31 − 12	becomes	31	− 10	−	2	=	19
2	58 − 24	becomes		−		−		=	
3	47 − 25	becomes		−		−		=	
4	66 − 31	becomes		−		−		=	
5	82 − 44	becomes		−		−		=	

34 − 23
becomes
34 − 20 − 3 = 11

Space Turns

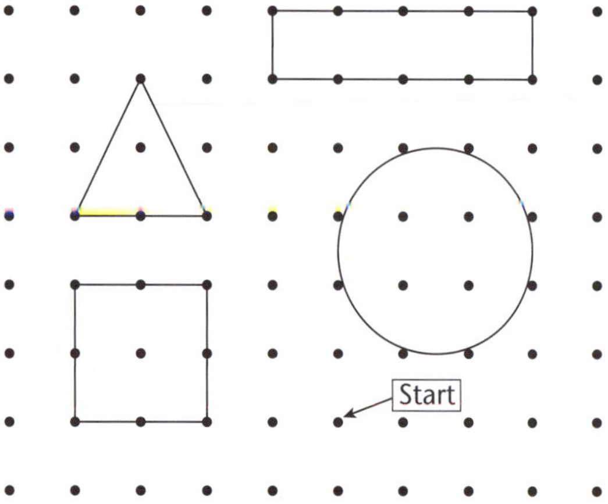

Follow the directions to draw a path on the map. You may need to turn the page so you can see which way is clockwise.

1 Start at the starting point.

2 Move 2 spaces towards the top of the page.

3 Make a quarter turn anticlockwise and move 3 spaces.

4 Make a quarter turn clockwise and move 1 space.

5 Make a quarter turn clockwise and move 4 spaces.

6 In which shape does the path end?

Number and Algebra

SET 3 Multiplication/arrays (5s and 10s)

Write and solve a number sentence to describe the arrays.

1 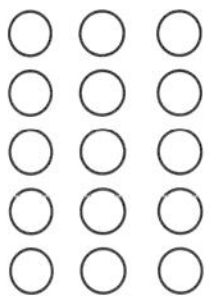

5 × ☐ = ☐

2 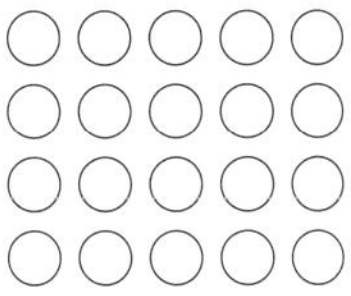

4 × ☐ = ☐

3 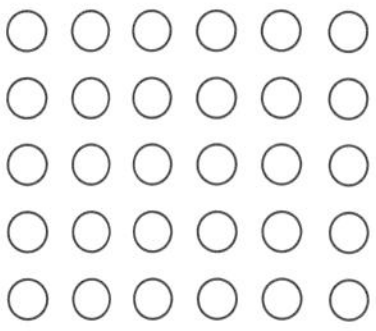

5 × ☐ = ☐

4 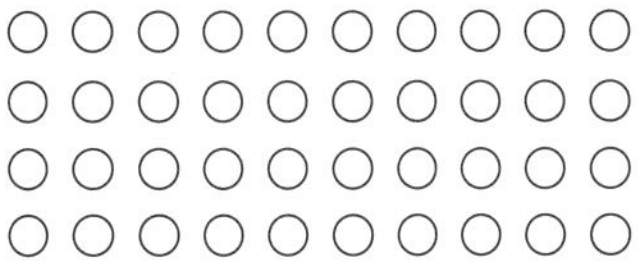

☐ × ☐ = ☐

5 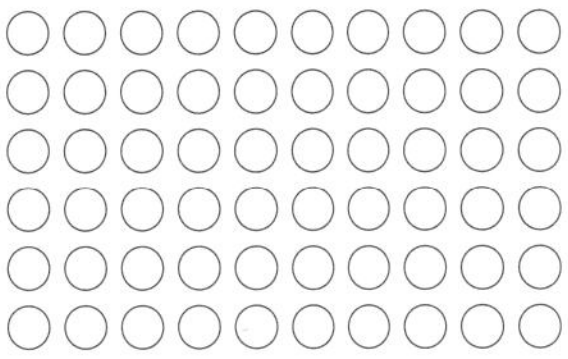

☐ × ☐ = ☐

SET 4 Extension

1 Which is bigger: $\frac{1}{3}$ or $\frac{1}{2}$?

2 How many weeks in a year?

3 10 + 6 + 5

4 ☐ − 4 = 44

5 How many 50c cakes can I buy with $2.50?

6 18 plus 6

7 How many months in 2 years?

8 How many days in 3 school weeks?

9 If the big hand is on the 12 and the small hand is on the 3, what is the time?

10 125c = $ ☐ . ☐ ☐

11 $5.50 take away 25c

12 What is the total number of sides for this group of shapes?

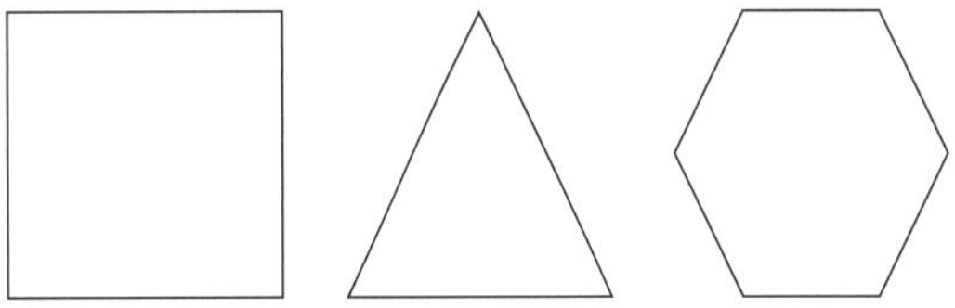

Measurement Time

Draw a line to match the label to the clock face.

half past 2

10 past 11

8 o'clock

20 past 12

25 past 3

Number and Algebra

SET 1 Basic

1. 14 − 4 = ____
2. 10 − 6 = ____
3. 4 + 4 = ____
4. 16 − 6 = ____
5. 20 − 5 = ____
6. 10 + 8 = ____
7. Two less than 7 = ____
8. Sum of 13 and 6 = ____
9. 2 × 7 = ____

10

Which number comes between 16 and 18?

____

SET 2 The division symbol (÷ means how many)

Use the array to answer the questions.

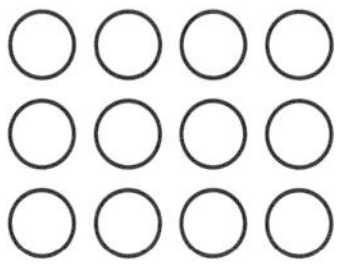

1. 12 ÷ 3 = ____
2. 12 ÷ 4 = ____

3. 15 ÷ 5 = ____
4. 15 ÷ 3 = ____

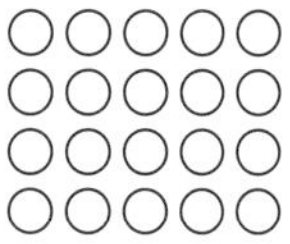

5. 20 ÷ 4 = ____
6. 20 ÷ 5 = ____

Number and Algebra Halves, quarters and eighths

1. Draw circles to break the group of 8 into halves.

2. Draw circles to break the group of 16 into quarters.

3. Draw circles to break the group of 16 into eighths.

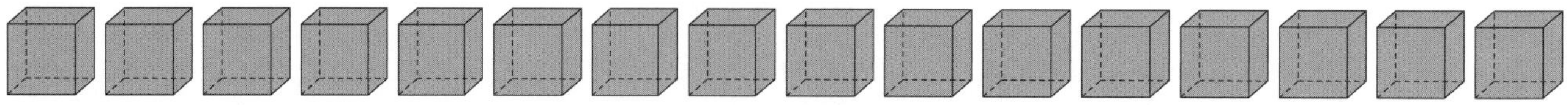

Number and Algebra

SET 3 Trading in addition to 99

Complete the additions.

1

	Tens	Ones
	3	6
+	2	5

2

	Tens	Ones
	3	8
+	1	4

3

	Tens	Ones
	4	7
+	2	6

4

	Tens	Ones
	5	7
+	1	4

5

	Tens	Ones
	4	4
+	3	8

6

	Tens	Ones
	6	5
+	1	7

7

	Tens	Ones
	4	9
+	2	4

8

	Tens	Ones
	5	8
+	2	5

SET 4 Extension

1 6 + 4 + 6

2 How many seconds in one minute?

3 Cost of a $3 cake and a $4 tart

4

$$\begin{array}{r} \$1.50 \\ +\ 1.25 \\ \hline \\ \hline \end{array}$$

5

$$\begin{array}{r} \$2.25 \\ +\ 1.75 \\ \hline \\ \hline \end{array}$$

6 One train ticket is 25c. How much for 3?

7 Nine divided by three

8 Which is larger: 1 kg or 100 g?

9 How many minutes in one hour?

10 65 + 25

11 ☐ divided by 4 = 5

12 How much is 9 worth in 393?

13 7 + ☐ = 13

14 How many legs do 3 cats have?

Measurement The metre

Colour the correct box to describe the lengths.

	Objects	Less than 1 metre	About 1 metre	More than 1 metre
1	The length of your desk			
2	The height of a door			
3	The width of your bedroom			
4	The height of your teacher			
5	The height of your desk			

Number and Algebra

SET 1 Basic

		− 5
1	5	
2	7	
3	8	
4	10	
5	15	
6	20	

7 10 + 5 = ☐

8 17 − 5 = ☐

9 12 − 5 = ☐

10 If Alicia lost 5 marbles out of her packet of 17, how many does she have left? ☐

SET 2 Split strategy

Use the split strategy to add these numbers mentally.

1 43 + 24 = ☐

2 52 + 35 = ☐

3 36 + 23 = ☐

4 17 + 31 = ☐

5 53 + 36 = ☐

6 41 + 44 = ☐

7 62 + 25 = ☐

8 13 + 55 = ☐

Space Symmetry

Complete the symmetrical shapes.

1

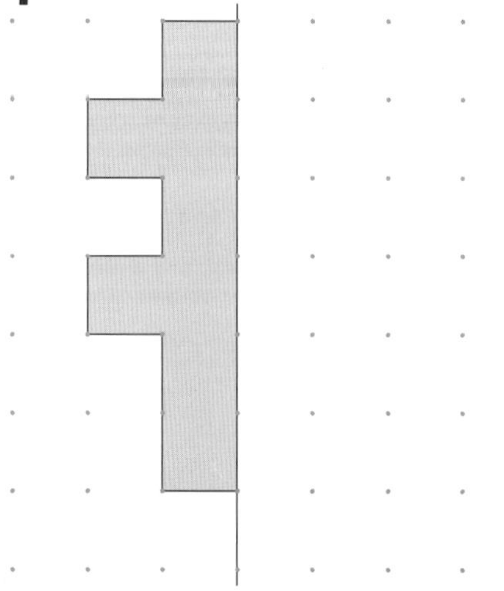

2

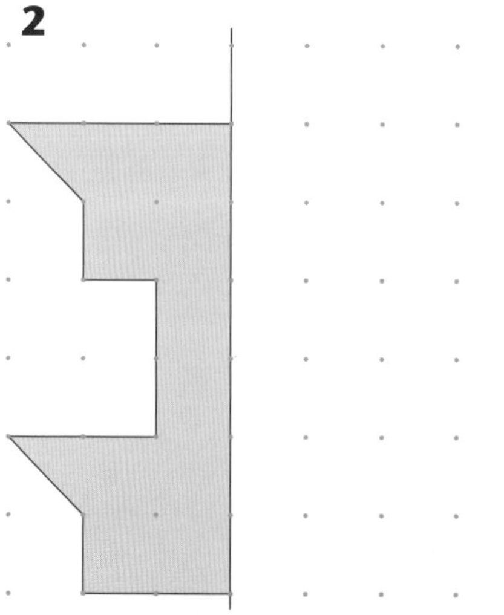

3

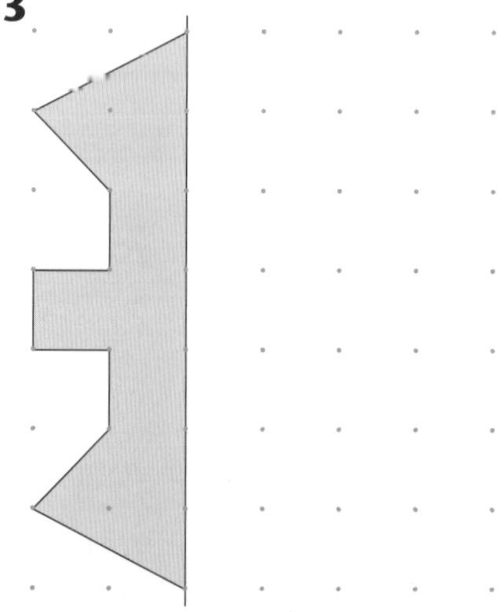

Number and Algebra

SET 3 Think board/multiplication problems

1	Problem: How much would it cost Jack to buy 3 books?	Number sentence and answer ☐ ☐ ☐ = ☐
	Model ? 4 | 4 | 4	 $4
2	Problem: How much would it cost Maddie to buy 4 staplers?	Number sentence and answer ☐ ☐ ☐ = ☐
	Model ? 6 | 6 | 6 | 6	 $6

SET 4 Extension

1 Write 29 in words.

____

2 How many minutes in half an hour?

3 2 + 7 + 8 =

4 How many days in December?

5 25 − 7

6 What are two shapes that have 4 sides?

7 How many days in 3 weeks?

8 18 + ☐ = 30

9 What date is New Year's Eve?

10 Which is heavier: 2 kg of feathers or 1 kg of lead?

11 How much are two combs at $2.50 each?

12 Value of 8 in 3086

13 Draw a pattern of a square, then a triangle, then a pentagon, then a circle.

Space Turns

Write $\frac{1}{4}$, $\frac{1}{2}$ or $\frac{3}{4}$ after deciding how each card has been turned clockwise.

1 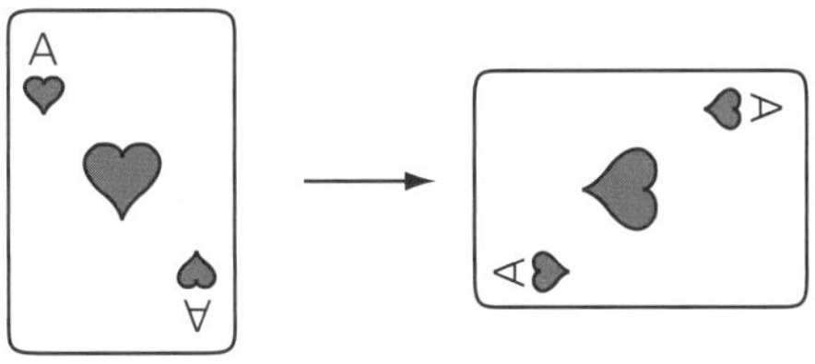 ☐

2 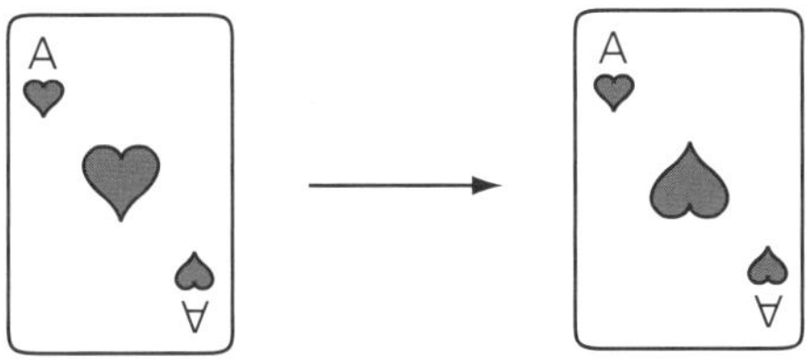 ☐

3 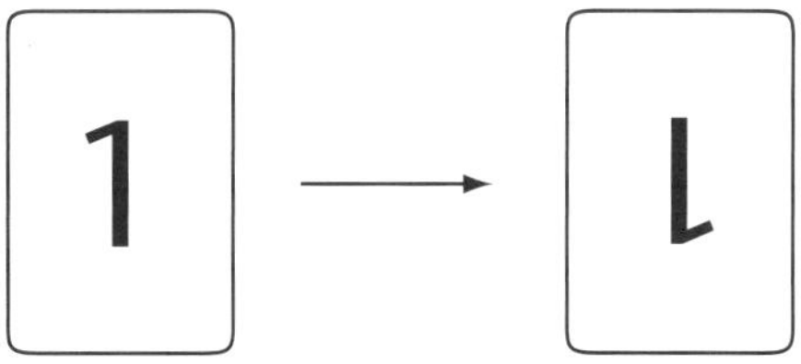 ☐

4 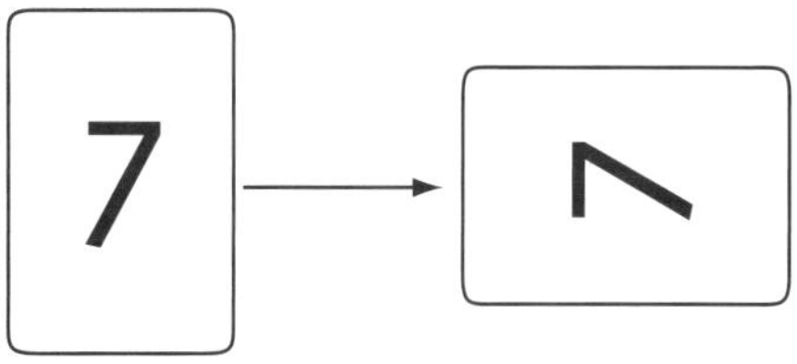 ☐

Number and Algebra

SET 1 Basic

Use the array to answer the multiplications.

1	2 × 5	☐	**6**	2 × 10	☐
2	5 × 5	☐	**7**	5 × 10	☐
3	3 × 5	☐	**8**	3 × 10	☐
4	4 × 5	☐	**9**	4 × 10	☐
5	6 × 5	☐	**10**	6 × 10	☐

SET 2 Rounding to 100

Round each number to the nearest 100.

1 291 ≈ ______ **3** 320 ≈ ______ **5** 505 ≈ ______

2 375 ≈ ______ **4** 595 ≈ ______ **6** 385 ≈ ______

Round each number to the nearest 100 to make adding easier.

7 199 + 199 ≈ ☐ **10** 196 + 108 ≈ ☐

8 98 + 295 ≈ ☐ **11** 497 + 215 ≈ ☐

9 303 + 199 ≈ ☐ **12** 509 + 196 ≈ ☐

Space Following directions

1 Draw a path using a red pencil from X to K.

2 Draw a path from Y to Z using a blue pencil.

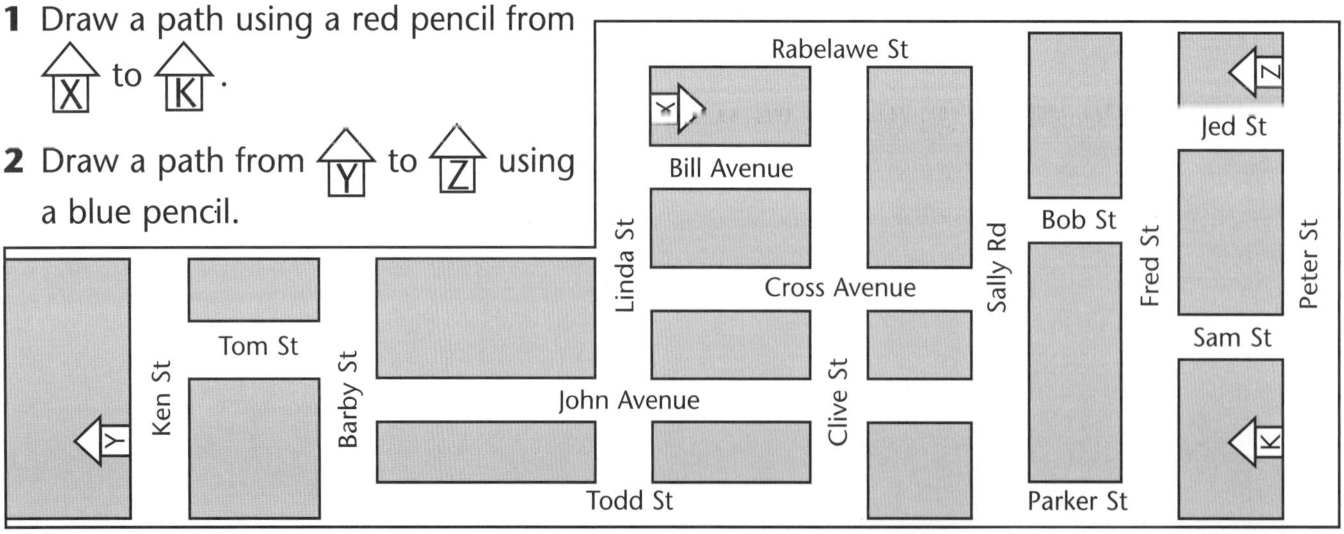

Number and Algebra

SET 3 Shopkeeper's method

Calculate how much has to be added on to the smaller number to complete the subtractions.

1	90 – 76	think (76)	+		+		=		
2	70 – 53	think (53)	+		+		=		
3	50 – 32	think (32)	+		+		=		
4	60 – 49	think (49)	+		+		=		

SET 4 Extension

1 $\frac{1}{4}$ of 40

2 How many legs on one spider?

3 How many legs on three spiders?

4 $2 \times \square = 40$

5 Share $100 among four people.

6 How many days in a leap year?

7 13, 15, 17, ____, 21

8 21 + 12 + 9

9 What number comes after 99?

10 Is 99 odd or even?

11 There are 30 people in Kate's class but 7 are away today. How many people are there today?

12 How much money is there below?

$____ . ____ ____

Measurement Time

Write an activity that could be measured using the time units in the list.

1	**Seconds**	
2	**Minutes**	
3	**Hours**	
4	**Days**	
5	**Weeks**	
6	**Months**	

Number and Algebra

SET 1 Basic

		+ 8
1	1	
2	2	
3	5	
4	7	
5	10	
6	11	

7 [20] − [8] = [ ]

8 [12] + [8] = [ ]

9 [6] + [8] = [ ]

10 What is half of 14? [ ]

SET 2 Multiplication/arrays (4s)

Write and solve a number sentence to describe the arrays.

1 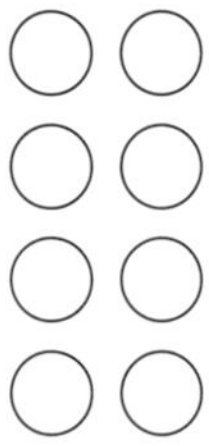

[4] × [ ] = [ ]

2 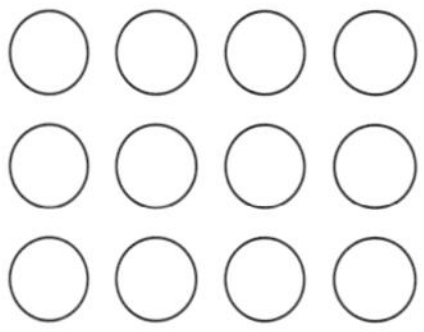

[3] × [ ] = [ ]

3 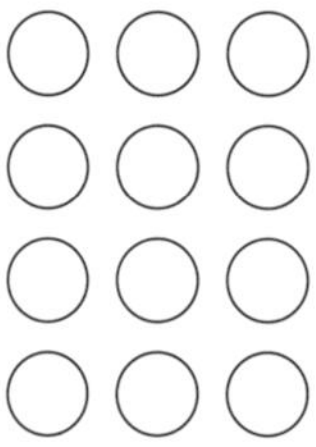

[ ] × [ ] = [ ]

4 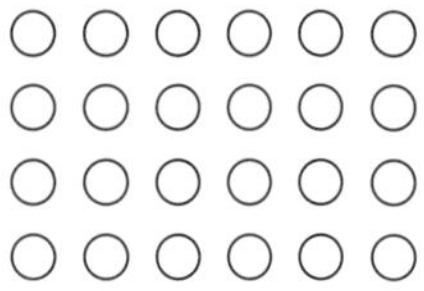

[ ] × [ ] = [ ]

5 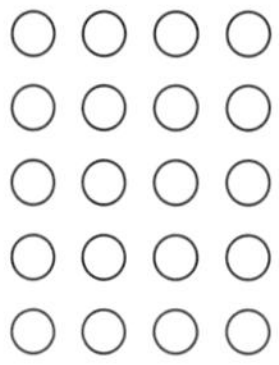

[ ] × [ ] = [ ]

Space Grid references

Put a dot on the grid for each grid reference given.

1 F2

2 F5

3 B5

4 B4

5 A4

6 A2

Join the dots in the order presented to form a shape.

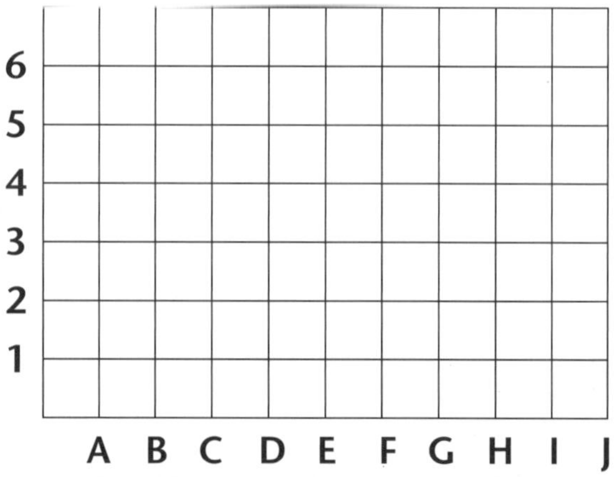

Number and Algebra

SET 3 Think board/division problems

1	Problem: Adnan divided a group of blocks into four equal groups. How many were in each group?	Number sentence and answer ☐ ☐ ☐ = ☐
	Model 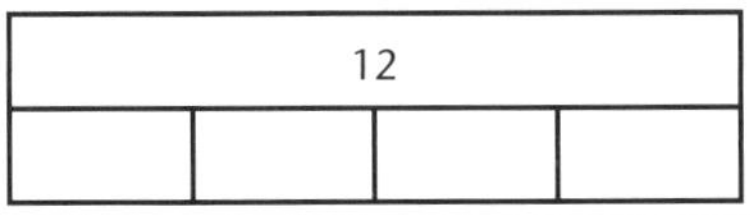	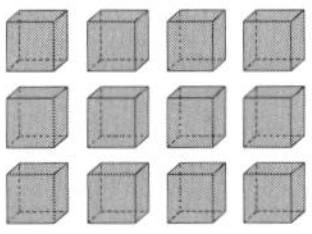
2	Problem: Cyra divided a group of blocks into 6 equal groups. How many were in each group?	Number sentence and answer ☐ ☐ ☐ = ☐
	Model 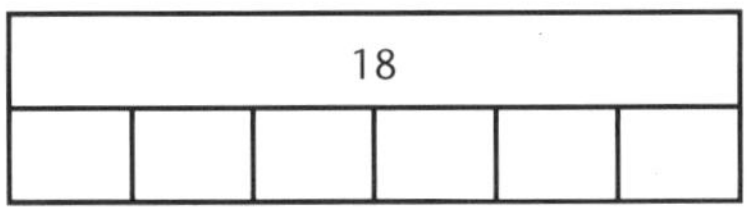	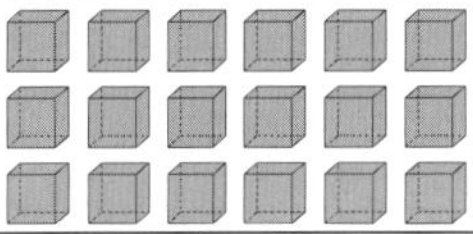

SET 4 Extension

1 4 + 4 + 4

2 $\frac{1}{4} = \square$ %

3 $\frac{1}{2} = \square$ %

4 7 tens plus 13 ones

5 Which number sentence has the answer 21?

○ 15 + 8 ○ 30 − 8
○ 3 × 6 ○ 3 × 7

6 Mia is 5 and Jenna is twice her age. How old is Jenna?

7 8 + ☐ = 13

8 If two drinks cost $4, how much would one be?

9 5 into 25 goes ☐ times.

10 How many grams in a kilogram?

11 Round 293 to the nearest 100.

12 Half of 90

Statistics Interpreting data

Y2 Favourite Sports

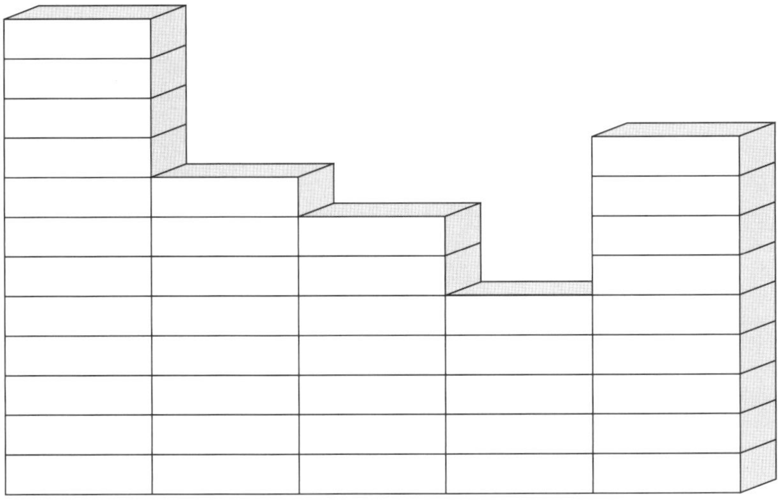

Soccer Netball Cricket Rugby Tennis

1 How many children played soccer?

____

2 How many children played cricket?

____

3 Write a question about the graph. ______________________

____

____

____

Number and Algebra

SET 1 Basic

1. 11 + 6 = ☐
2. 12 − 5 = ☐
3. 10 + 4 = ☐
4. 8 − 3 = ☐
5. 18 − 7 = ☐
6. 6 − 3 = ☐
7. ☐ + 3 = 10
8. 19 − 11 = ☐
9. 1 ten and 9 ones = ☐

10

Liam played 16 games of football and lost 7. How many did he win? ☐

SET 2 Compensation strategy

Use this strategy to complete these problems. The first one has been done for you.

1	25 + 19	becomes	25 + 20 − 1 = 44
2	27 +18	becomes	27 + ☐ − ☐ = ☐
3	32 + 28	becomes	32 + ☐ − ☐ = ☐
4	26 + 29	becomes	26 + ☐ − ☐ = ☐
5	34 + 27	becomes	34 + ☐ − ☐ = ☐

Space Turns

Circle one of the shapes in the box which shows what the original shape would look like after it has been turned clockwise.

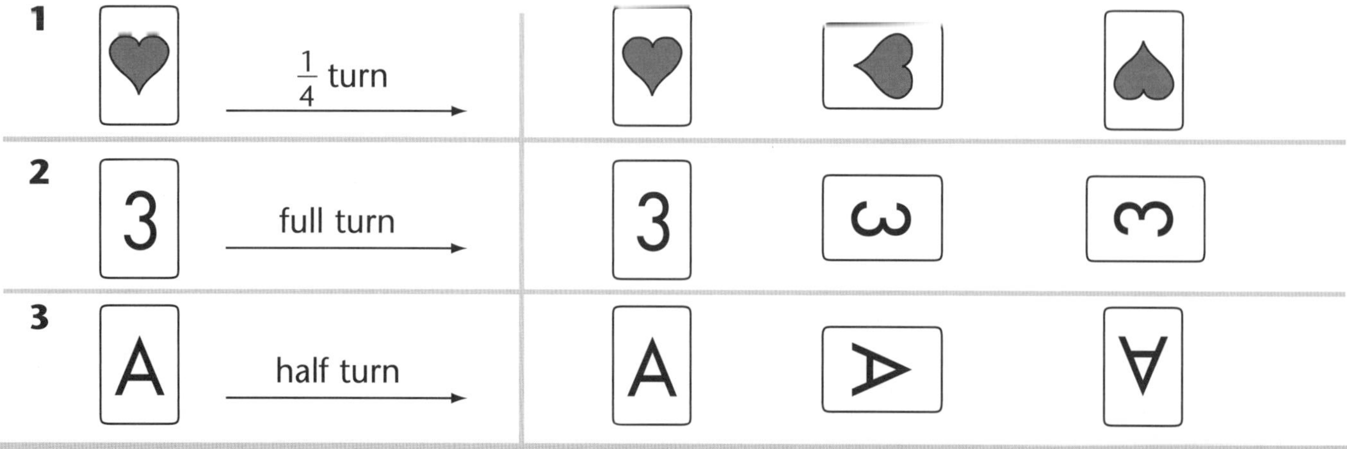

Number and Algebra

SET 3 Constant term input-output machines

Discover the rule then complete the grids.

1

Input	9	10	11	12	13	14
Output	10		12			

2

Input	7	8	9	10	11	12
Output	12			15		

3

Input	10	11	12	13	14	15
Ouput	12	13				

4

Input	10	11	12	13	14	15
Ouptut			9			12

Add the missing numbers to the pattern.

5

1	3		7			13	15

6

20			26	28	30		

SET 4 Extension

1 How much do 2 kg of potatoes cost at $2.50 per kilo?

2 3 calculators at $4 each

3 20 + 35

4 ☐ + 12 = 24

5 Half of 80

6 Is 25 an odd number?

7 Kelly saved $5 a month for 10 months. How much did she save?

8 4 lots of 6

9 Is December a summer month?

10 Place the numbers 6, 2, 1 and 7 into the shapes so that each line adds to 13.

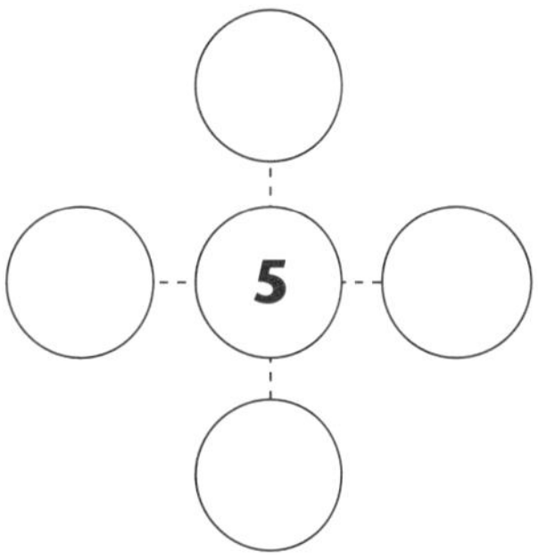

Measurement Area

1 Colour the largest shape red.

2 Colour the smallest shape blue.

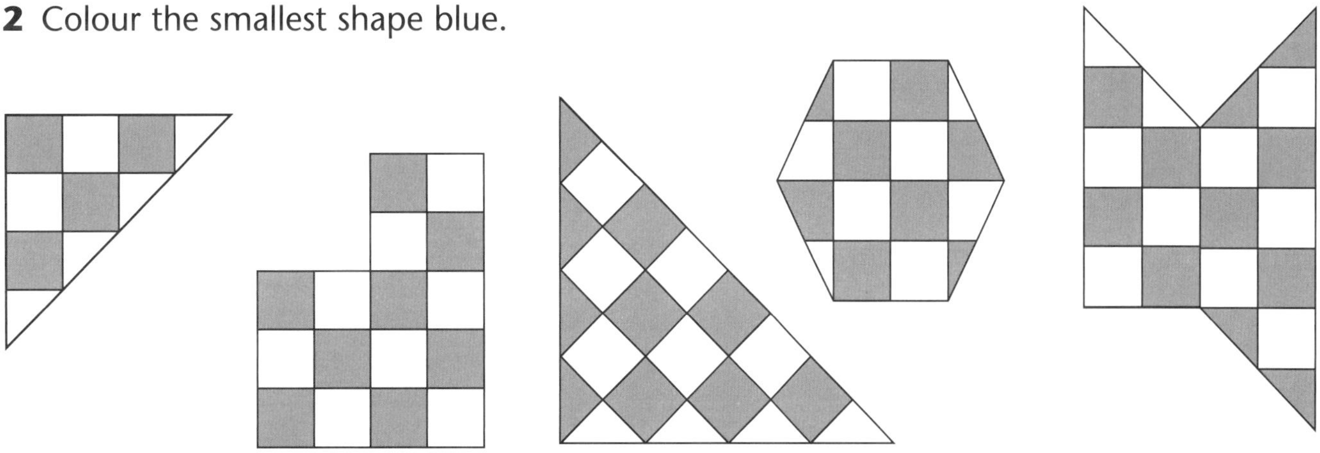

Number and Algebra

SET 1 Basic

		+ 10
1	1	
2	2	
3	3	
4	4	
5	5	
6	6	

7 19 − 10 = ☐

8 10 + 10 = ☐

9 20 − 10 = ☐

10 Diab had 6 red lollies and 6 blue lollies. How many does he have altogether? ☐

SET 2 Division facts

Use the array to help you divide the numbers by 5.

1 25 divided by 5 = ☐

2 15 divided by 5 = ☐

3 50 divided by 5 = ☐

4 45 divided by 5 = ☐

5 35 divided by 5 = ☐

6 20 divided by 5 = ☐

Space Reflections and translations

Write **reflection**, **turn** or **translation** to describe how the shape has been moved.

1

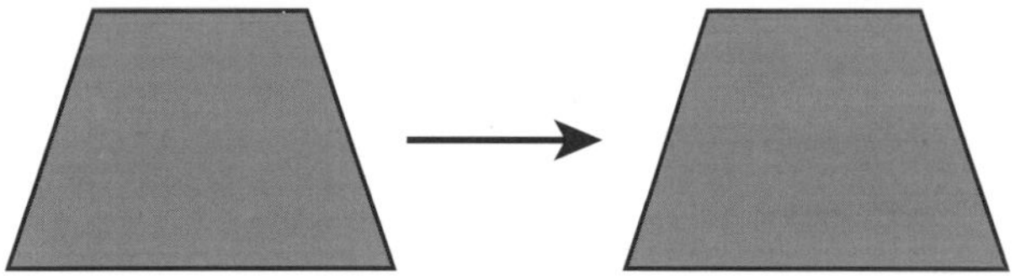

2

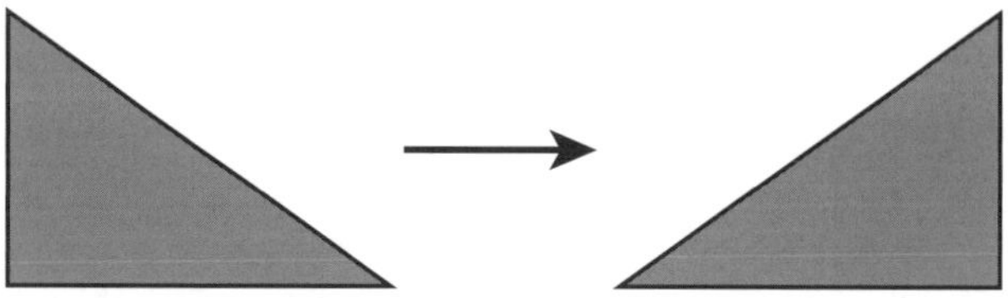

Number and Algebra

SET 3 Trading in subtraction

Complete the subtractions.

1

	Tens	Ones
	5	2
−	2	3

2

	Tens	Ones
	6	3
−	3	6

3

	Tens	Ones
	7	1
−	2	3

4

	Tens	Ones
	8	2
−	3	4

5

	Tens	Ones
	9	3
−	4	5

6

	Tens	Ones
	8	1
−	5	3

7

	Tens	Ones
	7	2
−	2	3

8

	Tens	Ones
	6	2
−	2	4

SET 4 Extension

1 10, 15, 20, ____, ____, 35

2 How many centimetres in 3 m?

3 How many sixes = 12?

4 Ian had 36 stickers.
How many groups of 6 can he make?

5 4 + 8 + 3 + 1

6 How many 50c coins make $3?

7 Is a prism a 2D shape?

8 Which is bigger: $\frac{1}{3}$ or $\frac{1}{4}$?

9 Show 9:30 on the clock below.

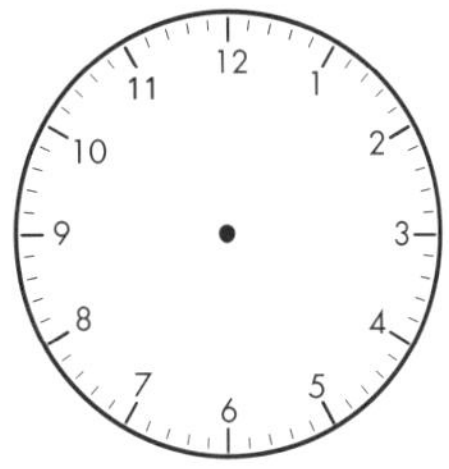

10 50 km − 19 km

11 Write the number 5 less than 92.

12 Is 63 odd or even?

Measurement Comparing measurement attributes

Draw arrows to show the length, width and height of the box.

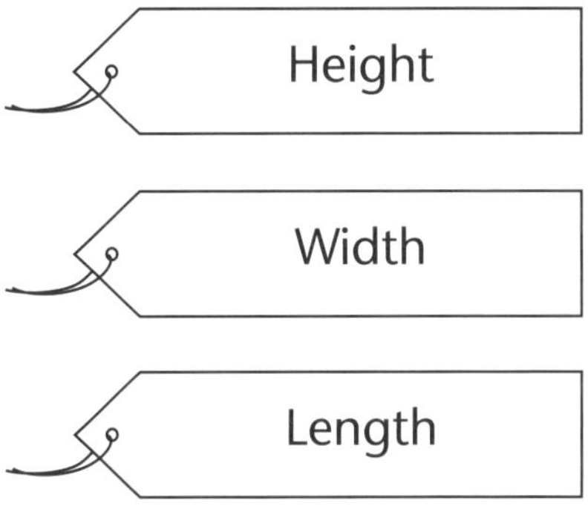

Number and Algebra

SET 1 Basic

1 11 − 2 = ___

2 6 − 3 = ___

3 10 + 8 = ___

4 4 + 4 = ___

5 8 − 2 = ___

6 15 − 7 = ___

7 One number after 16 = ___

8 Double 8

9 20c plus 40c

10 Jess made 15 cakes and sold 8. How many did she have left? ___

SET 2 Turnarounds

Supply the missing numbers in these turnarounds.

1 7 + ___ = 10   3 + ___ = 10

2 5 + ___ = 9   4 + ___ = 9

3 12 + ___ = 15   3 + ___ = 15

4 9 + 4 = ___   4 + ___ = 13

Space Maps

1 Draw trees in the park that is between Kelly Rd, Hodge St, High St and Five St.

2 The school is on the corner of Bardot and Five St. Put a cross where it is.

3 Draw a house which is between Tom St, Jack St, Jill St and Kelly Rd.

4 Draw Tim's house, which is on the corner of Log Rd and Jack St.

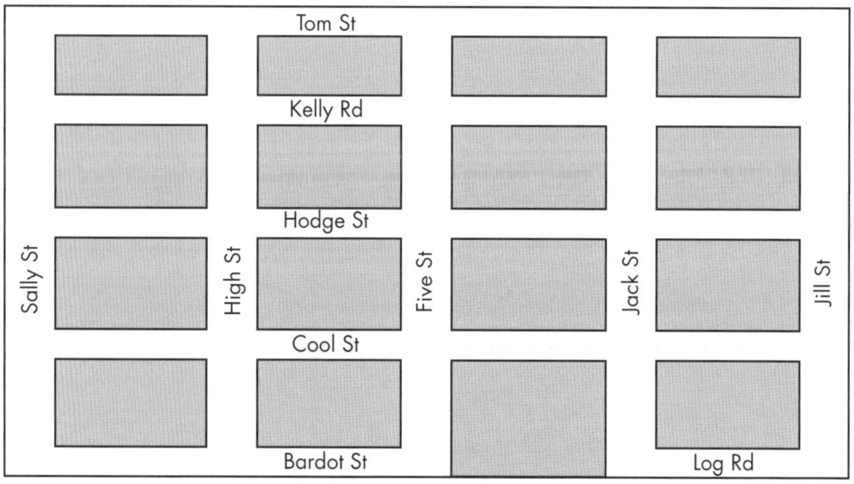

Number and Algebra

SET 3 Division

Use the number line to solve the divisions.

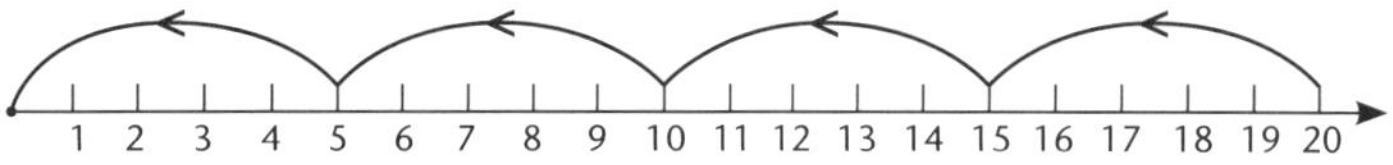

1 [14] ÷ [2] = [ ]

2 [20] ÷ [2] = [ ]

3 [6] ÷ [2] = [ ]

4 [18] ÷ [2] = [ ]

5 [20] ÷ [5] = [ ]

6 [15] ÷ [5] = [ ]

7 [10] ÷ [5] = [ ]

SET 4 Extension

1 100 – 83

2 Half of 60

3 \$2 = [ ] cents

4 Sum of 16 and 3

5 Jackson has 5 boxes each containing 10 apples. How many apples does he have altogether?

6 3 + 3 + 2

7 9 + [ ] = 16

8 25, 50, ____, 100

9 Circle the largest number.
101, 111, 001

10 Enlarge the shape by doubling its dimensions.

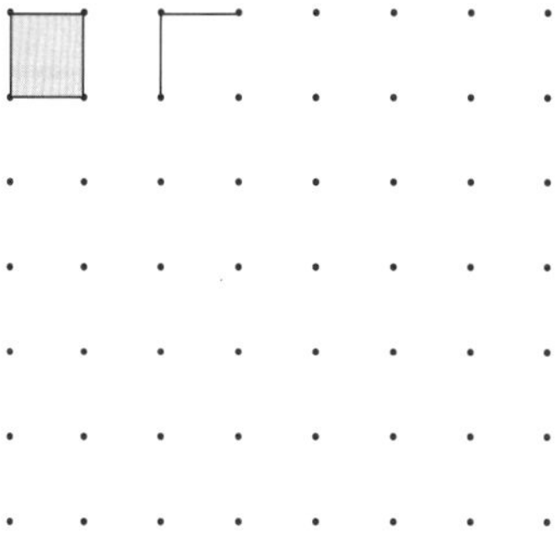

Statistics Survey

Class 2T surveyed their class to find out what they do in the afternoon.

Computer	𝍸 II
Sport	𝍸 𝍸 I
TV	𝍸
Music	III
Play	𝍸 I

Colour the blocks to make a graph of the data.

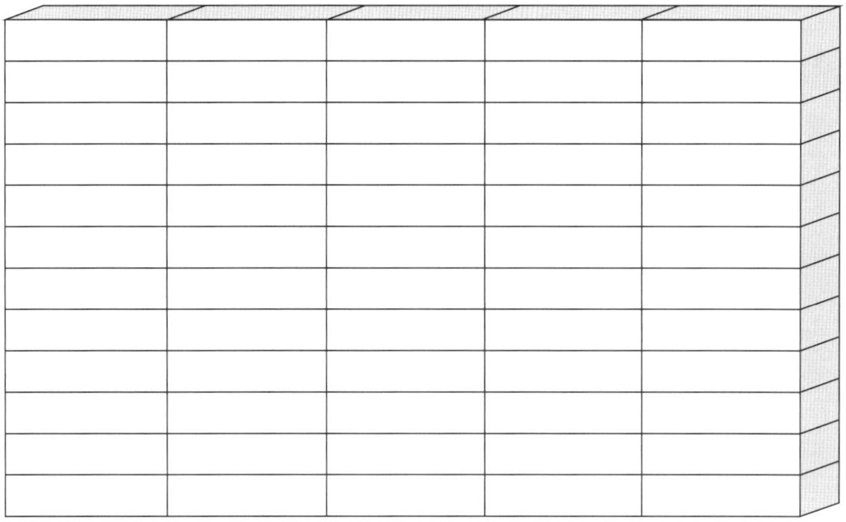

Computer Sport TV Music Play

UNIT 31

Number and Algebra

SET 1 Basic

		+ 9
1	1	
2	2	
3	3	
4	4	
5	9	
6	11	

7 19 − 9 = ☐

8 10 + 9 = ☐

9 8 + 9 = ☐

10 Jayne bought two tops at $9 each. How much did she spend? ☐

SET 2 Multiplication arrays (3s)/strategies

Use the array of footballs to answer the multiplications.

1 3 × 3 = ☐

2 5 × 3 = ☐

3 7 × 3 = ☐

4 6 × 3 = ☐

5 4 × 3 = ☐

Number and Algebra Repeating patterns

1 Starting with the colour green, colour the circles to make a repeating pattern of 3, 2. Use red for the other colour.

2 Starting with the colour blue, colour the squares to make a repeating pattern of 2, 1, 2. Use yellow for the 1 and red for the other colour.

Number and Algebra

SET 3 Constant term number patterns

Complete the number patterns then write the rules.

1	13	15	17	19				Rule ________
2	5	10	15	20				Rule ________
3	10	13	16	19				Rule ________
4	0	4	8	12				Rule ________
5	20	30	40	50				Rule ________

SET 4 Extension

1 Write the largest number you can using 1, 3, 4.

2 How many months in spring?

3 The value of 3 in 392

4 Halves in 2 wholes

5 What 2 coins make 70c?

6 Six lots of six = ☐

7 Write the number for 3 hundreds + 4 tens + 6 ones. ________

8 Add 2 tens to 302.

9 205, 207, ____, 211, ____

10 Share 35 among 5.

11 Write the smallest number you can using 2, 4, 1.

12 Half of 32

13 Write the number 10 more than 73.

Measurement The kilogram

Find and list things measured in kilograms.

Items	Kilograms

Number and Algebra

SET 1 Basic

1 12 − 6 = ☐

2 14 + 4 = ☐

3 19 − 8 = ☐

4 20 − 11 = ☐

5 15 + 5 = ☐

6 10 − 6 = ☐

7 3 × 5 = ☐

8 2 × 9 = ☐

9 Double 8 = ☐

10 Amy had two 20c pieces and 3 $1 coins. How much money did she have? ☐

SET 2 Money transactions

Show which coins would be needed to make each amount.

	Amount					
1	$0.85					
2	$1.95					
3	$1.35					
4	$2.65					
5	$2.90					

Space 2D revision

Colour match the labels.

1 I am a shape that has 3 corners/vertices.

2 I am a shape that has 4 equal sides and angles.

3 I am a shape that has 6 sides.

4 I am a shape that has 8 sides.

5 I am like a square but two of my sides are longer.

Octagon

Triangle

Square

Hexagon

Rectangle

Number and Algebra

SET 3 Remainders

Write a division sentence to describe each array.

1 There are ☐ groups of 5 with ☐ remainders.

2 There are ☐ groups of 4 with ☐ remainders.

SET 4 Extension

1 7 more than 25

2 2 weeks + 3 days = ☐ days

3 200 – 5 =

4 Double 21.

5 50% of 100

6 How many groups of 30 can go into 90?

7 How many minutes in $1\frac{1}{2}$ hours?

8 Round 839 to the nearest 10.

9 Sum of 6 and 27

10 Multiply 5 and 3.

11 Put these numbers in order from smallest to largest.
293, 339, 103, 406, 239, 101

Measurement Calendar

1 Complete the calendar for March

March

Sun	Mon	Tues	Wed	Thur	Fri	Sat
			1	2	3	4
5						

2 Answer the questions.

a What day of the week is March 5? ____________

b What day of the week is March 17? ____________

c What day of the week is March 22? ____________

d What is the date of the second Thursday in March? ____________

e What is the date of the second Monday in March? ____________

Hundreds chart

1	2	3	4	5	6	7	8	9	10
11	12	13	14	15	16	17	18	19	20
21	22	23	24	25	26	27	28	29	30
31	32	33	34	35	36	37	38	39	40
41	42	43	44	45	46	47	48	49	50
51	52	53	54	55	56	57	58	59	60
61	62	63	64	65	66	67	68	69	70
71	72	73	74	75	76	77	78	79	80
81	82	83	84	85	86	87	88	89	90
91	92	93	94	95	96	97	98	99	100

Addition grid

+	0	1	2	3	4	5	6	7	8	9	10	11	12
0	0	1	2	3	4	5	6	7	8	9	10	11	12
1	1	2	3	4	5	6	7	8	9	10	11	12	13
2	2	3	4	5	6	7	8	9	10	11	12	13	14
3	3	4	5	6	7	8	9	10	11	12	13	14	15
4	4	5	6	7	8	9	10	11	12	13	14	15	16
5	5	6	7	8	9	10	11	12	13	14	15	16	17
6	6	7	8	9	10	11	12	13	14	15	16	17	18
7	7	8	9	10	11	12	13	14	15	16	17	18	19
8	8	9	10	11	12	13	14	15	16	17	18	19	20
9	9	10	11	12	13	14	15	16	17	18	19	20	21
10	10	11	12	13	14	15	16	17	18	19	20	21	22
11	11	12	13	14	15	16	17	18	19	20	21	22	23
12	12	13	14	15	16	17	18	19	20	21	22	23	24

UNIT 1 Number and Algebra

SET 1

1 6
2 5
3 4
4 3
5 7
6 9
7 3
8 2
9 6
10 7

SET 2

1 2 tens 4 ones
2 3 tens 2 ones
3 5 tens 6 ones
4 1 tens 2 ones
5 2 tens 8 ones
6 9 tens 9 ones
7 62
8 75

SET 3

1
a 5 **b** 3
2
a 6 **c** 7
b 9 **d** 10
3
a 7 **c** 8
b 10 **d** 11
4
a 10 **c** 11
b 13 **d** 14

SET 4

1 2 tens + 5 ones
2 4 + 6
3 4
4 39
5 16
6 82
7 $10
8 9 + 9
9
a
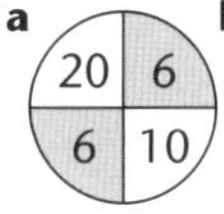

b
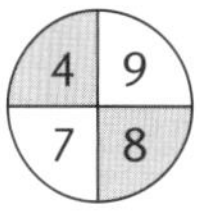

Space

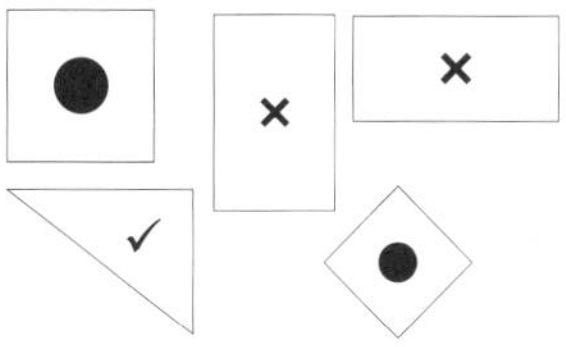

Measurement

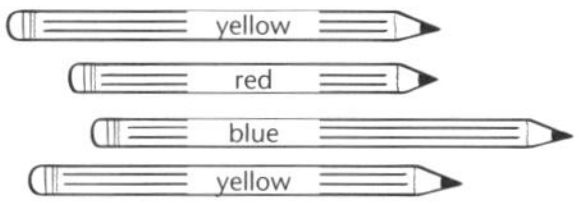

UNIT 2 Number and Algebra

SET 1

1 3
2 5
3 7
4 8
5 6
6 9
7 8
8 5
9 3
10 5

SET 2

1 False
2 False
3 True
4 True
5 True
6 False
7 False
8 True

SET 3

1 14
2 16
3 19
4 22
5 10
6 12
7 18
8 22
9 13

SET 4

1 16
2 126
3 2
4 5
5 36
6 5 tens 6 ones
7 $10
8 20
9 4th, 5th
10 Hands on, some examples could be:
a 3 + 2
b 1 + 4
c 2 + 3

Space

1 Car
2 Piggy bank
3 Clock
4 Toy soldier
5 Male doll

Measurement

1 2 o'clock
2 4 o'clock
3 9 o'clock
4 11 o'clock

UNIT 3 Number and Algebra

SET 1

1 6
2 5
3 1
4 7
5 4
6 2
7 4
8 4
9 8
10 8 lollies

SET 2

1 9
2 9
3 10
4 9
5 10
6 13
7 11
8 12
9 14
10 17
11 16

SET 3

1 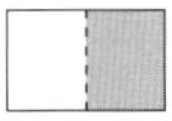
2 
3 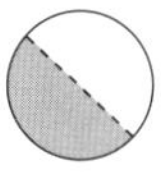
4 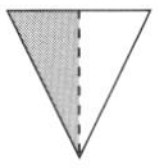
5 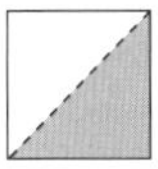
6 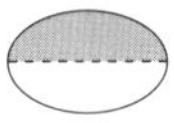
7 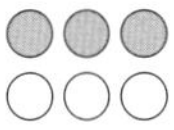
8 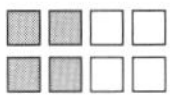

SET 4

1
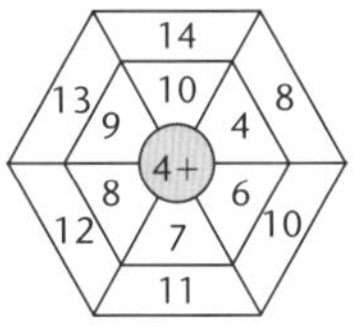

2 winter
3 17
4 15
5 10
6 0
7 31
8 1 tens 8 ones
9 7
10 215, 512, 125
11 April

Number and Algebra

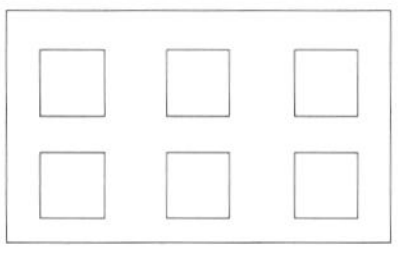
double 3 is 6
half of 6 is 3

Measurement

1 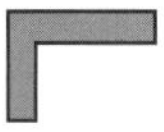
2 
3 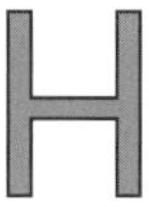
4 
5 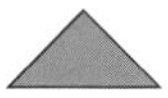
6 

UNIT 4 Number and Algebra

SET 1

1 6
2 8
3 3
4 8
5 5
6 8
7 2
8 1
9 5
10 6

SET 2

1 5
2 2
3 3
4 25
5 12
6 10
7 7
8 2
9 4

SET 3

1

2

SET 4

1 seven
2 10
3 12
4 Even
5 Equal in length
6 14
7 7
8 8
9 30
10 October
11 $3.80

Space

1 Rectangular prism
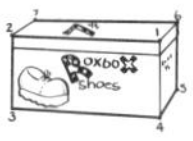
2 Triangular prism

3 Cube
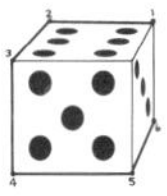

Measurement

1

5 o'clock
2

Half past 3
3

Half past 9
4

Half past 7
5

Half past 10

Answers

UNIT 5 Number and Algebra

SET 1

1 1
2 2
3 3
4 4
5 5
6 6
7 5
8 7
9 7
10 7

SET 2

1 8, 10
2 20, 25
3 26, 30
4 16, 20
5 35, 40
6 34, 38
7 55, 50
8 54, 52

SET 3

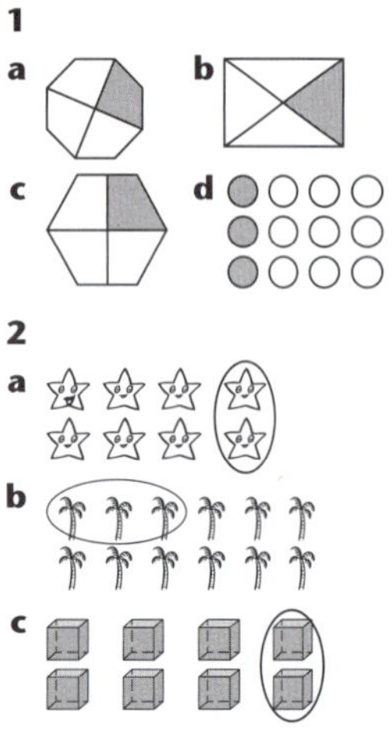

SET 4

1 31
2 10
3 [3] tens [6] ones
4 December
5 13
6 10
7 21
8 
9 8 − 2 = 6
10 [3] hundreds [5] tens [6] ones
11 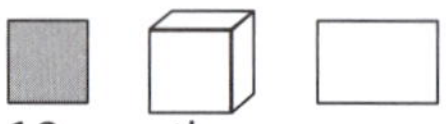
12 12 months

Measurement

1 7 shoes, almost 8
2 5 shoe boxes, almost 6

Statistics

1

Cat	IIII	Bird	IIII
Dog	卌 I	Lizard	I
Fish	III		

2

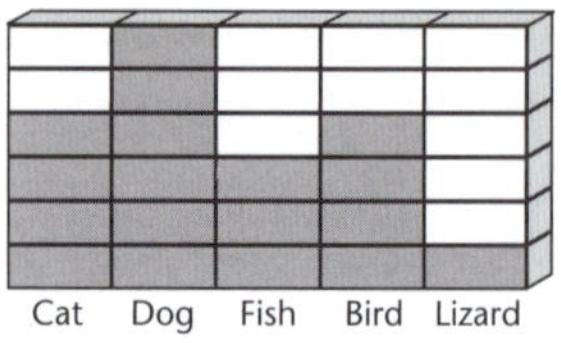

UNIT 6 Number and Algebra

SET 1

1 10
2 9
3 9
4 5
5 12
6 5
7 7
8 5
9 2
10 2

SET 2

1 [10] + [10] + [10] = [30]
2 [10] + [10] + [10] + [10] = [40]
3 [10] + [10] + [10] + [10] + [10] = [50]
4 [5] + [5] + [5] + [5] = [20]

SET 3

1 4
2 4
3 3
4 3
5 6
6 6
7 5
8 5
9 10
10 10

SET 4

1 Numerous responses
2 26, 28
3 6
4 Autumn
5 [7] + [3] = [10]
6 9
7 3
8 15c
9 Equal in weight
10 5c
11 20
12 25 April
13 6

Space

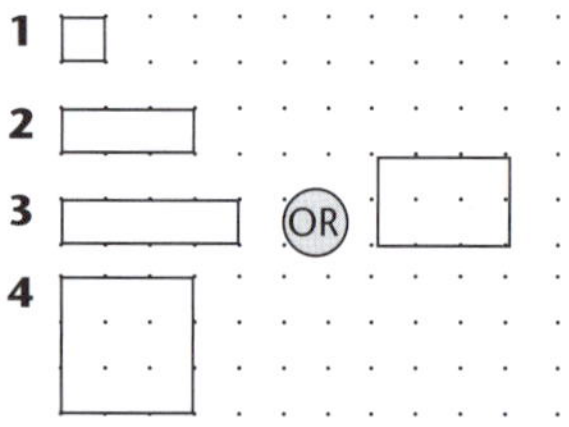

Measurement

UNIT 7 Number and Algebra

SET 1

1 4
2 5
3 6
4 7
5 8
6 9
7 5
8 10
9 8
10 4

SET 2

1 4
2 4
3 6
4 4
5 2
6 5
7 5
8 6
9 4

SET 3

1 3 groups of 3
2 3 groups of 4
3 5 groups of 10
4 6 groups of 2

SET 4

1 10, 13, 23, 99
2 14
3 6
4 14
5 $20
6 yes
7 $1.65
8 6
9 20
10 8
11 Hands on

Space

Hands on

Measurement

1 Hands on
2 Hands on

UNIT 8 Number and Algebra

SET 1

1 5
2 4
3 1
4 4
5 8
6 5
7 6
8 8
9 8
10 5

SET 2

1 [4] groups of [3] = [12]
2 [4] groups of [4] = [16]
3 [5] groups of [2] = [10]
4 [3] groups of [5] = [15]

SET 3

1 2
2 3
3 2
4 4
5 3
6 6
7 4
8 4
9 3
10 3

SET 4

1 Three hundred and ten
2 9
3 4
4 15
5 8
6 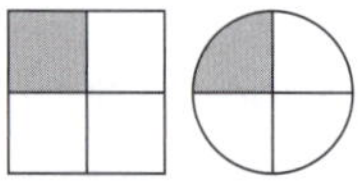
7 31 days
8 2 halves
9 299
10 100
11 April
12

+	3	6	9	12	15	18	21
3	6	9	12	15	18	21	24

Statistics

Marble	Tally
1	卌 I
2	III
3	IIII
4	I
5	III

Measurement

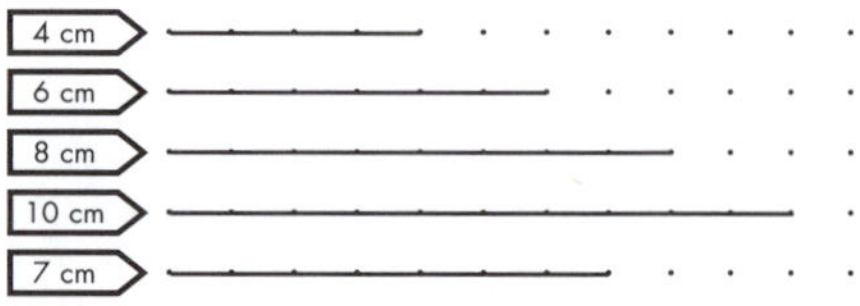

UNIT 9 Number and Algebra

SET 1

1 1
2 2
3 3
4 4
5 5
6 6
7 11
8 9
9 7
10 5

SET 2

1 77
2 89
3 77
4 86
5 79
6 99
7 68
8 59

SET 3

1 [3] groups of [4] = [12]
2 [4] groups of [10] = [40]
3 [5] groups of [3] = [15]
4 [4] groups of [5] = [20]

SET 4

1 9
2 40
3 6
4 6
5 24, 26, 28
6 [2] tens [7] ones
7 100
8 12
9 
10 8
11 December
12 17
13 69

Space

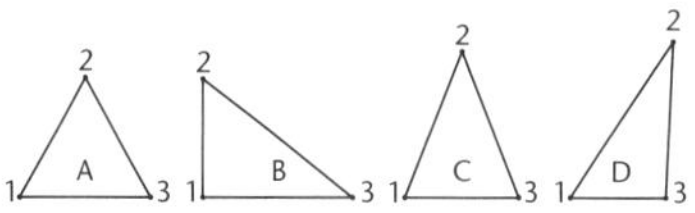

Measurement

1 2 hours
2 5 hours
3 45 minutes
4 30 minutes

UNIT 10 Number and Algebra

SET 1

1 5
2 12
3 7
4 5
5 6
6 8
7 8
8 4
9 5
10 6 pets

SET 2

1 [7] tens [5]
2 [3] tens [9]

	Number	Hundreds	Tens	Ones
3	324	III	II	IIII
4	681	𝍸 I	𝍸 III	I
5	653	𝍸 I	𝍸	III
6	702	𝍸 II		II
7	993	𝍸 IIII	𝍸 IIII	III

SET 3

1 10
2 10
3 20
4 30
5 20
6 40
7 40
8 30
9 90
10 50

SET 4

1 36
2 21
3 13 + 2
4 3 months
5 15
6 24
7 80c
8 23
9 2, 4, 6, 8, 10, 12, 14
10 24
11 26 December
12 13
13 21
14 14

Number and Algebra

1 30
2 51

Space

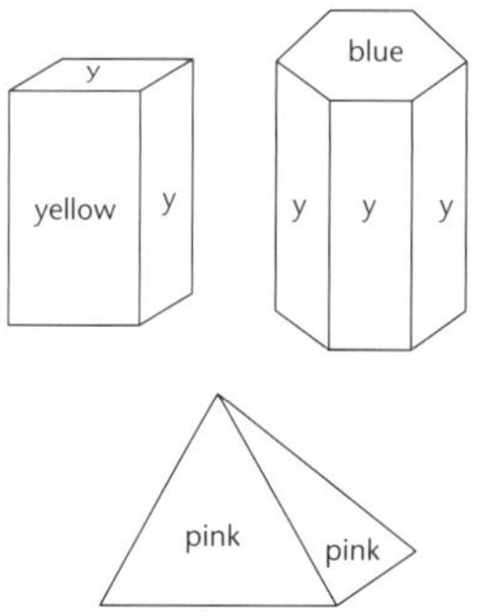

UNIT 11 Number and Algebra

SET 1

1 5
2 6
3 7
4 8
5 9
6 10
7 15
8 10
9 10
10 14 laps

SET 2

1 5 × [2] = [10]
2 3 × [4] = [12]
3 5 × [5] = [25]
4 3 × [10] = [30]

SET 3

1 [14] + [12] = [26]
2 [28] + [49] = [77]
3 [31] + [27] = [58]

SET 4

1 52, 62, 72, 82, 92, 102
2 4
3 15
4 $\frac{1}{2}$ or $\frac{2}{4}$
5 12
6 6
7 50c
8 30
9 34, 44, 54, 64, 74, 84
10 90
11 100 cm
12 2 dozen
13 10, 13, 22, 23, 43, 83

Space

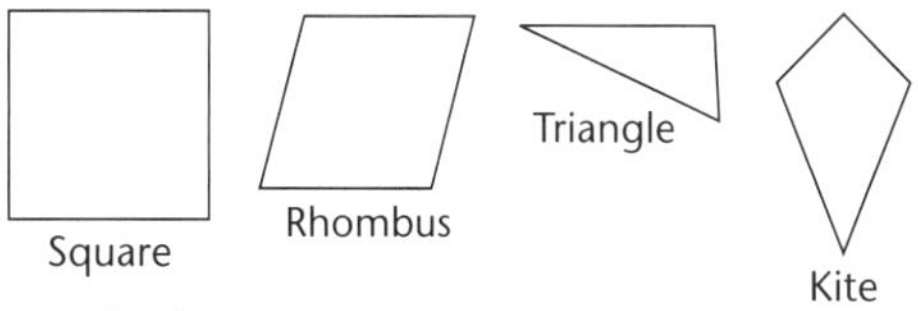

Statistics

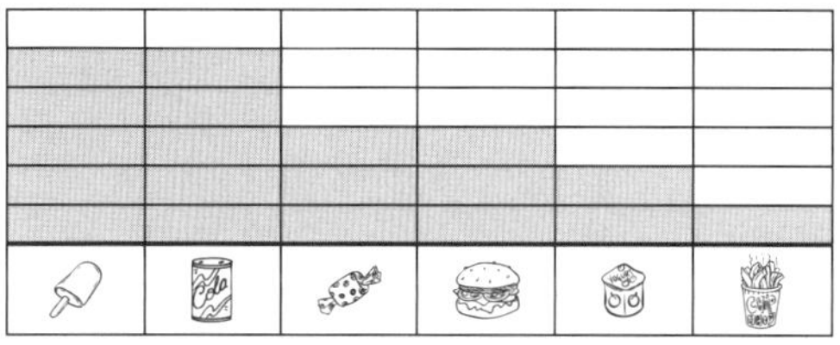

UNIT 12 Number and Algebra

SET 1

1 16
2 12
3 6
4 6
5 6
6 10
7 5
8 15
9 12
10 15 cows

SET 2

1 [2] = 6 popsticks
2 [3] = 4 popsticks
3 [4] = 3 popsticks
4 [6] = 2 popsticks
5 [12] = 1 popstick

SET 3

1 28 lollies
2 Yes
3 15 counters
4 7 birds

SET 4

1 Winter
2 6
3 57
4 110 cents
5 30
6 6
7 5
8 76
9 30
10 40, 44, 48, 52, 56
11 31 days
12 55
13 8 legs
14 88

Number and Algebra

1 8
2 12
3 16
4 6
5 7
6 10

Measurement

Hands on

Answers

UNIT 13 Number and Algebra

SET 1

1 6
2 8
3 10
4 14
5 16
6 18
7 13
8 1
9 4
10 15

SET 2

2 hundreds 6 tens 5 ones
9 hundreds 8 tens 5 ones
3 hundreds 7 tens 3 ones
6 hundreds 4 tens 9 ones

SET 3

1 18
2 19
3 24
4 26
5 40
6 41
7 70
8 72
9 80
10 85

SET 4

1 31
2 9
3 83
4 3 sides
5 21 sides
6 24 hours
7 September
8 48
9 
10 April
11 $1.50
12 46
13 74
14 3rd

Number and Algebra

1 24
2 31

Space

1 
2 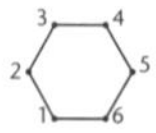
3 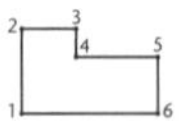

UNIT 14 Number and Algebra

SET 1

1 8
2 4
3 11
4 12
5 6
6 4
7 8
8 6
9 8
10 12

SET 2

1 20
2 43
3 10
4 12
5 27
6 04
7 57
8 25

SET 3

1 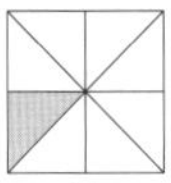
1/8
2 
1/4
3 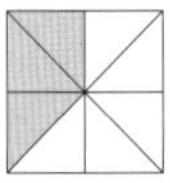
3/8
4 3/8

SET 4

1 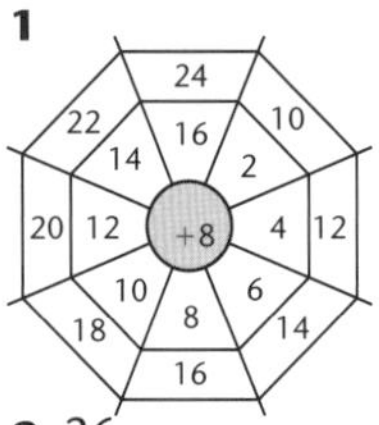

2 26
3 odd
4 302
5 67
6 24
7 20
8 60
9 36
10 $3.50
11 9 tens 8 ones
12 

Space

7 Cat and dog

Measurement

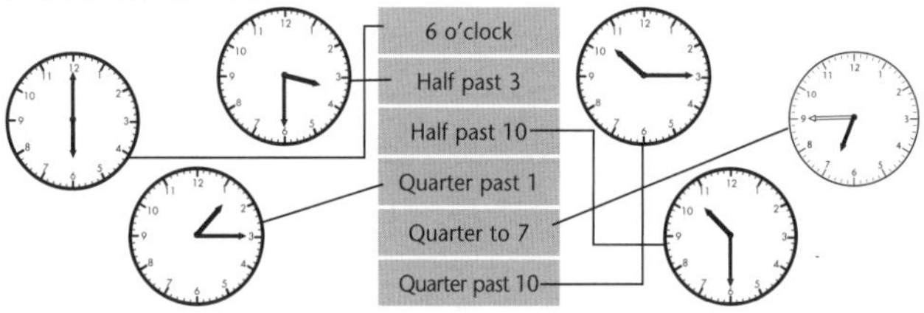

UNIT 15 Number and Algebra

SET 1

1 7
2 9
3 10
4 12
5 15
6 20
7 9
8 6
9 7
10 6

SET 2

1 4
2 6
3 2
4 8
5 10
6 14
7 12
8 16

SET 3

1 12
2 7
3 14
4 4 + 3 = 7
5 + 2 = 7
6 + 1 = 7

SET 4

1 12
2 78
3 12am
4 23
5 Summer
6 22
7 5 lollies each
8 7 tens 8 ones
9 38
10 12
11 18
12 $\frac{1}{2}$
13 48
14 Even

Space

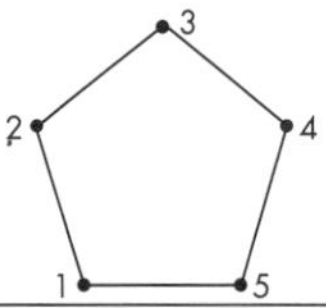
Name Pentagon
Sides 5 Corners 5

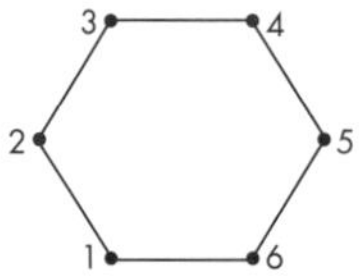
Name Hexagon
Sides 6 Corners 6

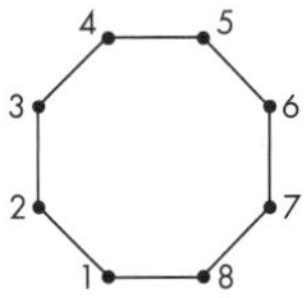
Name Octagon
Sides 8 Corners 8

Measurement

1 7 marbles
2 13 marbles
3 6 marbles
4 Yes

UNIT 16 Number and Algebra

SET 1

1 8
2 2
3 10
4 13
5 15
6 10
7 13
8 6
9 5
10 10

SET 2

1 34 + 10 + 3 = 47
2 36 + 20 + 2 = 58
3 24 + 20 + 3 = 47
4 43 + 20 + 4 = 67
5 52 + 20 + 4 = 76

SET 3

1 4
2 4
3 5

SET 4

1

+	1	2	3	4	10	11
9	10	11	12	13	19	20

2 75
3 365 days
4 12 legs
5 26
6 6 months
7 140
8 10
9 50
10 18 sides
11 12
12 9
13 $335

Measurement

1 7 cm
2 9 cm
3 5 cm
4 10 cm
5 8 cm
6 4 cm

Statistics

1 Blond
2 Black
3 3 students
4 1 student

UNIT 17 Number and Algebra

SET 1

1 0
2 2
3 3
4 5
5 7
6 12
7 10
8 13
9 15
10 15

SET 2

1 [3] × [2] = [6]
[2] × [3] = [6]

2 [4] × [3] = [12]
[3] × [4] = [12]

3 [3] × [5] = [15]
[5] × [3] = [15]

SET 3

1

12	13	14	15	16	17	18
10	10	10	10	10	10	10
22	23	24	25	26	27	28

2

14	24	34	44	54	64	74
6	6	6	6	6	6	6
20	30	40	50	60	70	80

3

12	22	32	42	52	62	72
5	5	5	5	5	5	5
17	27	37	47	57	67	77

4

24	34	44	54	64	74	84
7	7	7	7	7	7	7
31	41	51	61	71	81	91

SET 4

1 14
2 2 days
3 $1.35
4 14
5 22
6 300, 400, 500, 600, 700
7 $1.80
8 $2.50
9 23
10 50
11 8, 6, 4, 2
12 13
13 

Number and Algebra

1 10
2 90

Space

1 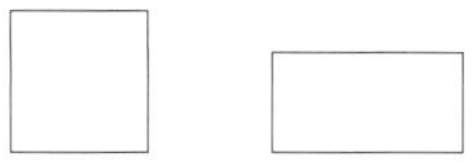

2 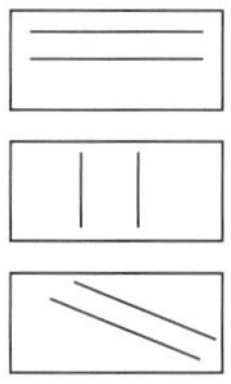

UNIT 18 Number and Algebra

SET 1

1 6
2 6
3 7
4 5
5 5
6 10
7 12
8 no
9 8
10 13

SET 2

1 0, 2, 4, 6, 8, 10, 12, 14, 16, 18
2 0, 5, 10, 15, 20, 25, 30, 35, 40
3 0, 10, 20, 30, 40, 50, 60, 70, 80, 90
4 0, 3, 6, 9, 12, 15, 18, 21, 24, 27

SET 3

1 11
2 9
3 11
4 12
5 16
6 15
7 11
8 14
9 14
10 12

SET 4

1 Even
2 7
3 2
4 16
5 11
6 48
7 4:30
8 $1.50
9 9 tens 5 ones
10 330
11 Monday
12 2 dozen
13 40

Geometry

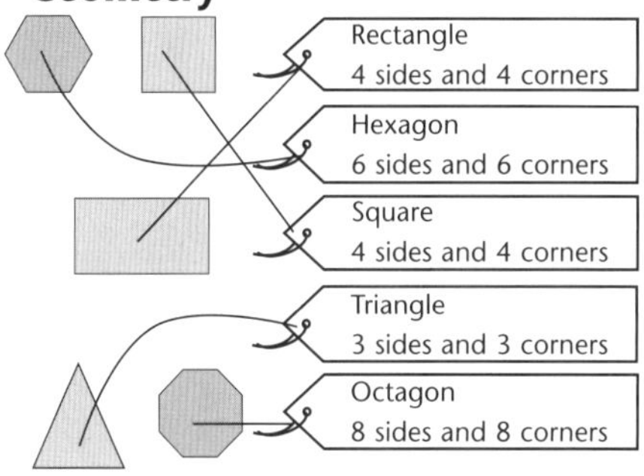

Statistics

Colour 8 marbles blue.
Colour 6 marbles red.
Colour 7 marbles green.

UNIT 19 Number and Algebra

SET 1

1 7
2 8
3 9
4 10
5 11
6 16
7 4
8 12
9 6
10 6

SET 2

1 45, 50, 55, 60, 65, 70, 75
2 30, 40, 50, 60, 70, 80, 90
3 12, 14, 16, 18, 20, 22, 24
4 8, 12, 16, 20, 24, 28, 32
5 95, 100, 105, 110, 115, 120, 125
6 46, 48, 50, 52, 54, 56, 58
7 110, 120, 130, 140, 150, 160, 170

SET 3

1 23
2 35
3 58
4 33
5 45

SET 4

1 17
2 972
3 76
4 $4.50
5 134
6 7 tens
7 3
8 Rule: going up by 10
9 100
10 97
11 77
12 6
13

–	10	13	16	19	20	50
3	7	10	13	16	17	47

Measurement

1 Friday
2 Friday
3 Tuesday
4 Friday
5 Thursday the 22nd November
6 Friday the 19th October

Space

Shape C

UNIT 20 Number and Algebra

SET 1

1 0
2 10
3 3
4 17
5 10
6 15
7 10
8 12
9 9
10 $11

SET 2

1 6
2 10
3 8
4 12
5 6
6 12
7 15

SET 3

1 2, 4, 6, 8, 10, 12, 14, 16
Rule: increasing by 2
2 20, 21, 22, 23, 24, 25, 26, 27
Rule: increasing by 1
3 40, 35, 30, 25, 20, 15, 10, 5
Rule: decreasing by 5
4 16, 26, 36, 46, 56, 66, 76, 86
Rule: increasing by 10
5 Hands on

SET 4

1 22
2 $13
3 21 days
4 8
5 90, 100, 110, 120, 130
Rule: Up by 10
6 56
7 5
8 24
9 40
10 678
11 876
12
a 32
b 41

Measurement

1 Colour the arrow.
2 Circle the ruler.
3 Tick the brush.

Space

a 

b 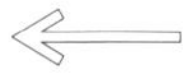

c 

Answers

UNIT 21 Number and Algebra

SET 1

1 1
2 2
3 6
4 11
5 12
6 16
7 10
8 10
9 14
10 $11

SET 2

1 38, 68, 86
2 22, 26, 32
3 106, 116, 126
4 333, 383, 393
5 74, 76, 176
6 77
7 78
8 27
9 33

SET 3

1 6
2 4
3 8
4 10
5 8
6 15
7 12
8 12

SET 4

1 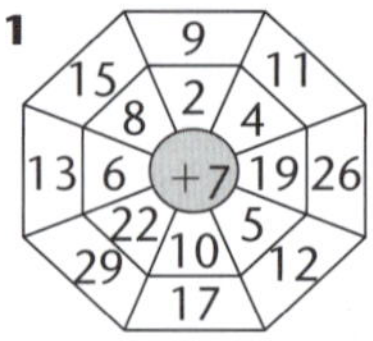

2 $2.50
3 90
4 10
5 even
6 13, 15, 17, 19, 21, 23
7 0
8 4
9 Answers will vary
10 Various responses

Number and Algebra

18 = 10 + 8, 12 + 6, 15 + 3
20 = 16 + 4, 13 + 7, 2 + 18
19 = 14 + 5, 3 + 16, 17 + 2

Statistics

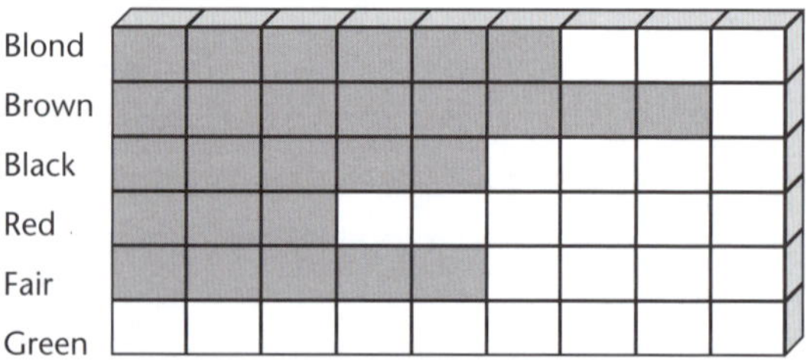

UNIT 22 Number and Algebra

SET 1

1 5
2 5
3 4
4 11
5 19
6 2
7 10
8 20
9 yes
10 17

SET 2

1, 2, 3, 4
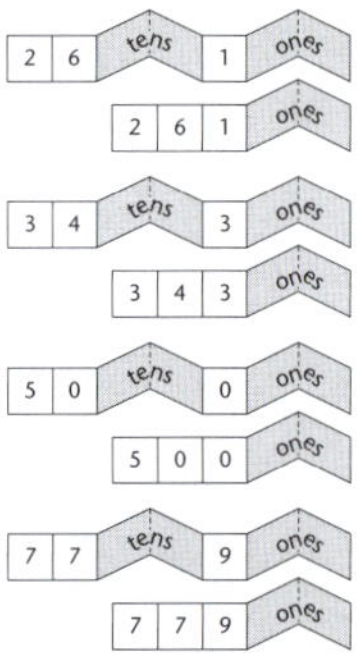

SET 3

1 (True) False
2 True (False)
3 (True) False
4 (True) False
5 True (False)

SET 4

1 18 + 7 = 25
2 19
3 20
4 $9.50
5 50
6 16 legs
7 70
8 2
9 6
10 4 halves
11 

Space

Answers may include:

It is a cube. It has 6 faces. Each face is a square. It has 12 edges and 8 corners.

Measurement

1 8 blocks
2 12 blocks
3 16 blocks
4 10 blocks

UNIT 23 Number and Algebra

SET 1

1 8
2 9
3 10
4 12
5 17
6 20
7 10
8 10
9 17
10 $5

SET 2

1 31 − 10 − 2 = 19
2 58 − 20 − 4 = 34
3 47 − 20 − 5 = 22
4 66 − 30 − 1 = 35
5 82 − 40 − 4 = 38

SET 3

1 5 × 3 = 15
2 4 × 5 = 20
3 5 × 6 = 30
4 4 × 10 = 40
5 6 × 10 = 60

SET 4

1 $\frac{1}{2}$
2 52 weeks
3 21
4 48
5 5
6 24
7 24
8 15
9 3 o'clock
10 $1.25
11 $5.25
12 13

Space

1–5 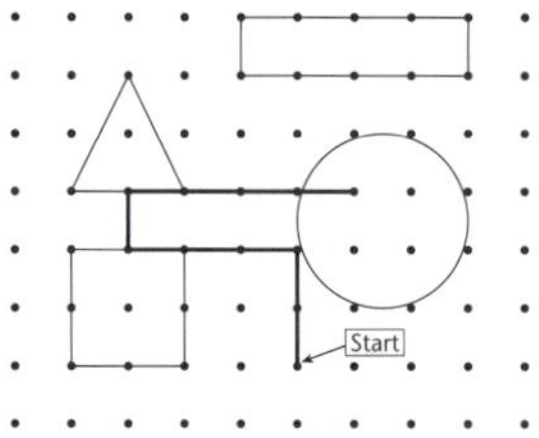

6 circle

Measurement

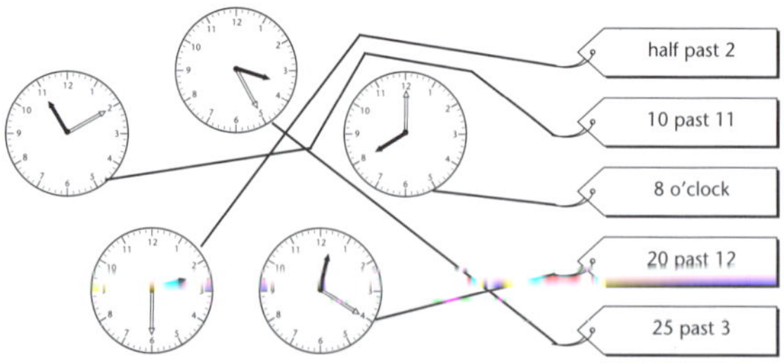

UNIT 24 Number and Algebra

SET 1

1 10
2 4
3 8
4 10
5 15
6 18
7 5
8 19
9 14
10 17

SET 2

1 12 ÷ 3 = 4
2 12 ÷ 4 = 3
3 15 ÷ 5 = 3
4 15 ÷ 3 = 5
5 20 ÷ 4 = 5
6 20 ÷ 5 = 4

SET 3

1 61
2 52
3 73
4 71
5 82
6 82
7 73
8 83

SET 4

1 16
2 60
3 $7
4 $2.75
5 $4
6 75c
7 3
8 1 kg
9 60 minutes
10 90
11 20 ÷ 4 = 5
12 90 or 9 tens
13 7 + 6 = 13
14 12 legs

Number and Algebra

1 Circle into 2 groups of 4.
2 Circle into 4 groups of 4.
3 Circle into 8 groups of 2.

Measurement

Hands on

UNIT 25 Number and Algebra

SET 1

1 0
2 2
3 3
4 5
5 10
6 15
7 15
8 12
9 7
10 12

SET 2

1 67
2 87
3 59
4 48
5 89
6 85
7 87
8 68

SET 3

1 [3] × [4] = [12]
2 [4] × [6] = [24]

SET 4

1 Twenty-nine
2 30
4 31
3 17
5 18
6 Square, rectangle etc.
7 21 days
8 12
9 31 December
10 2 kg feathers
11 $5.00
12 80 or 8 tens
13 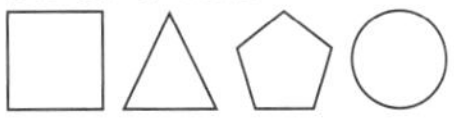

Space

1 2 3

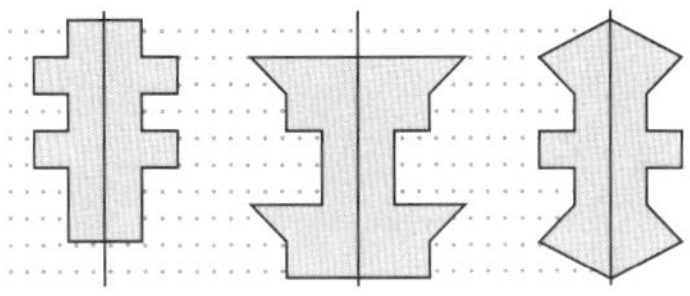

Space

1 $\frac{1}{4}$
3 $\frac{1}{2}$
2 $\frac{1}{2}$
4 $\frac{3}{4}$

UNIT 26 Number and Algebra

SET 1

1 10
2 25
3 15
4 20
5 30
6 20
7 50
8 30
9 40
10 60

SET 2

1 300
2 400
3 300
4 600
5 500
6 400
7 400
8 400
9 500
10 300
11 700
12 700

SET 3

1 [4] + [10] = [14]
2 [7] + [10] = [17]
3 [8] + [10] = [18]
4 [1] + [10] = [11]

SET 4

1 10
2 8
3 24
4 20
5 $25
6 366 days
7 13, 15, 17, <u>19</u>, 21
8 42
9 100
10 odd
11 23
12 $3.85

Space

Hands on

Measurement

Hands on

UNIT 27 Number and Algebra

SET 1

1 9
2 10
3 13
4 15
5 18
6 19
7 12
8 20
9 14
10 7

SET 2

1 [4] × [2] = [8]
2 [3] × [4] = [12]
3 [4] × [3] = [12]
4 [4] × [6] = [24]
5 [5] × [4] = [20]

SET 3

1 [12] ÷ [4] = [3]
2 [18] ÷ [6] = [3]

SET 4

1 12
2 25%
3 50%
4 83
5 3 x 7
6 Jenna is 10.
7 8 + 5 = 13
8 $2
9 5 times
10 1000 grams
11 300
12 45

Space

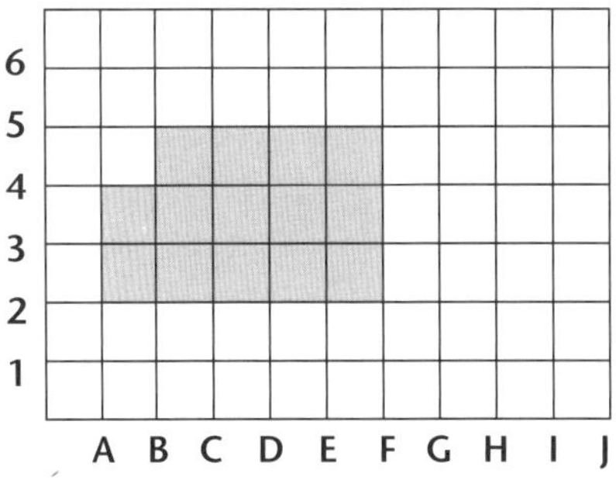

Statistics

1 12 children
2 7 children
3 Hands on

UNIT 28 Number and Algebra

SET 1

1 17
2 7
3 14
4 5
5 11
6 3
7 7
8 8
9 19
10 9 games

SET 2

1 25 + 20 − 1 = 44
2 27 + 20 − 2 = 45
3 32 + 30 − 2 = 60
4 26 + 30 − 1 = 55
5 34 + 30 − 3 = 61

SET 3

1

Input	9	10	11	12	13	14
Output	10	11	12	13	14	15

2

Input	7	8	9	10	11	12
Output	12	13	14	15	16	17

3

Input	10	11	12	13	14	15
Output	12	13	14	15	16	17

4

Input	10	11	12	13	14	15
Output	7	8	9	10	11	12

5

1	3	5	7	9	11	13	15

6

20	22	24	26	28	30	32	34

SET 4

1 $5
2 $12
3 55
4 12 + 12 = 24
5 40
6 Yes, odd
7 $50
8 24
9 Yes, in Australia
10 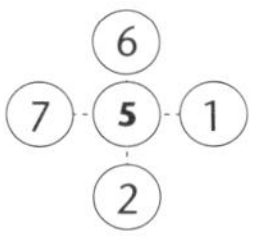

Space

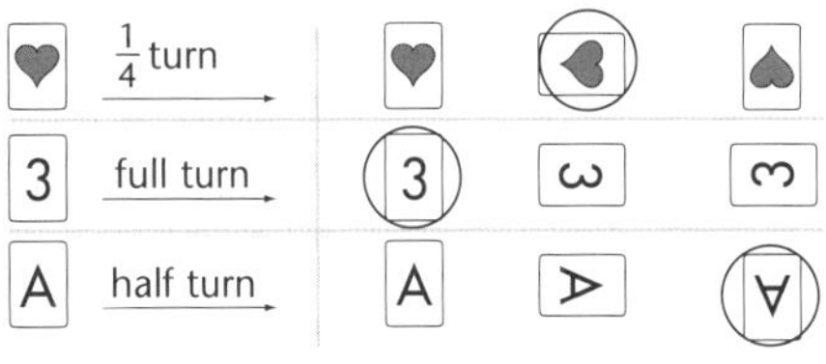

Measurement

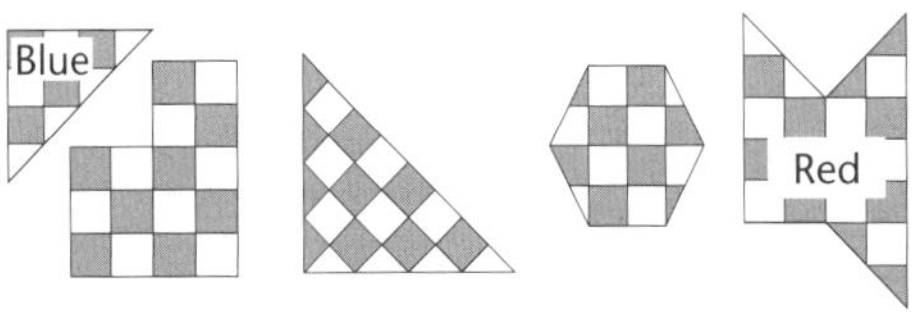

Answers

UNIT 29 Number and Algebra

SET 1

1 11
2 12
3 13
4 14
5 15
6 16
7 9
8 20
9 10
10 12

SET 2

1 5
2 3
3 10
4 9
5 7
6 4

SET 3

1 29
2 27
3 48
4 48
5 48
6 28
7 49
8 38

SET 4

1 10, 15, 20, 25, 30, 35
2 300 cm
3 2
4 6
5 16
6 6 coins
7 No
8 $\frac{1}{3}$
9

10 31 km
11 87
12 Odd

Space

1 translation
2 reflection

Measurement

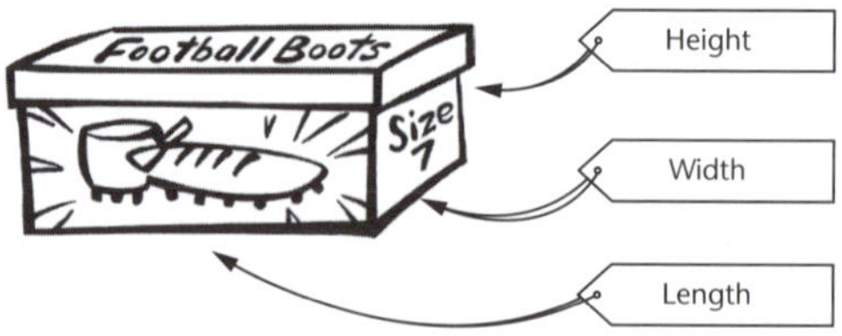

UNIT 30 Number and Algebra

SET 1

1 9
2 3
3 18
4 8
5 6
6 8
7 17
8 16
9 60c
10 7

SET 2

1 [7] + [3] = [10] [3] + [7] = [10]
2 [5] + [4] = [9] [4] + [5] = [9]
3 [12] + [3] = [15] [3] + [12] = [15]
4 [9] + [4] = [13] [4] + [9] = [13]

SET 3

1 7
2 10
3 3
4 9
5 4
6 3
7 2

SET 4

1 17
2 30
3 200 cents
4 19
5 50 apples
6 8
7 9 + 7 = 16
8 25, 50, 75, 100
9 101, (111), 001
10

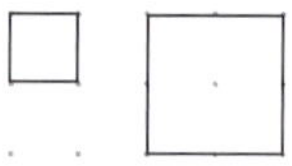

Space

Hands on

Statistics

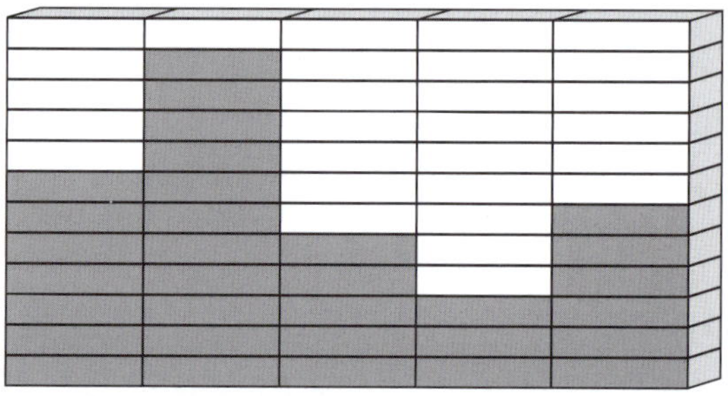

UNIT 31 Number and Algebra

SET 1

1 10
2 11
3 12
4 13
5 18
6 20
7 10
8 19
9 17
10 $18

SET 2

1 3 × 3 = [9]
2 5 × 3 = [15]
3 7 × 3 = [21]
4 6 × 3 = [18]
5 4 × 3 = [12]

SET 3

1 13, 15, 17, 19, 21, 23, 25
Increase by 2
2 5, 10, 15, 20, 25, 30, 35
Increase by 5
3 10, 13, 16, 19, 22, 25, 28
Increase by 3
4 0, 4, 8, 12, 16, 20, 24
Increase by 4
5 20, 30, 40, 50, 60, 70, 80
Increase by 10

SET 4

1 431
2 3 months
3 300 or 3 hundreds
4 4
5 50c and 20c
6 36
7 346
8 322
9 205, 207, 209, 211, 213
10 7 each
11 124
12 16
13 83

Number and Algebra

(G) (G) (G) (R) (R) (G) (G) (G) (R) (R) (G) (G) (G) (R) (R)

[B] [B] [Y] [R] [R] [B] [B] [Y] [R] [R] [B] [B] [Y] [R] [R]

Measurement

Hands on

UNIT 32 Number and Algebra

SET 1

1 6
2 18
3 11
4 9
5 20
6 4
7 15
8 18
9 16
10 $3.40

SET 2

Possible combination

	Amount					
1	$0.85	I	I	I	I	
2	$1.95	I		II	I	I
3	$1.35	I	I	I		I
4	$2.65	I	I		I	II
5	$2.90			II	I	II

SET 3

1 4 groups of 5 with [2] remainders
2 4 groups of 4 with [3] remainders

SET 4

1 32
2 17 days
3 195
4 42
5 50
6 3
7 90
8 840
9 33
10 15
11 101, 103, 239, 293, 339, 406

Space

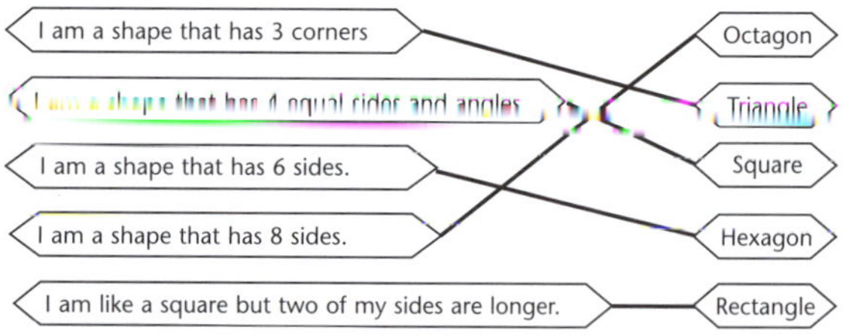

Measurement

1

Sun	Mon	Tue	Wed	Thur	Fri	Sat
			1	2	3	4
5	6	7	8	9	10	11
12	13	14	15	16	17	18
19	20	21	22	23	24	25
26	27	28	29	30	31	

2 a Sunday
b Friday
c Wednesday
d 9th
e 13th